APPLYING ANALYTICS

A Practical Introduction

APPLYING ANALYTICS

A Practical Introduction

E.S. Levine

CRC Press
Taylor & Francis Group
Boca Raton London New York

CRC Press is an imprint of the
Taylor & Francis Group, an **informa** business

A CHAPMAN & HALL BOOK

The views and opinions expressed in this book are those of the author and do not necessarily represent those of his current or former employers.

You can find links to the data sets used in this book at http://www.applyinganalytics.com

CRC Press
Taylor & Francis Group
6000 Broken Sound Parkway NW, Suite 300
Boca Raton, FL 33487-2742

First issued in paperback 2019

© 2014 by E.S. Levine
CRC Press is an imprint of Taylor & Francis Group, an Informa business

No claim to original U.S. Government works

ISBN-13: 978-1-4665-5718-5 (hbk)
ISBN-13: 978-0-367-37986-5 (pbk)

This book contains information obtained from authentic and highly regarded sources. Reasonable efforts have been made to publish reliable data and information, but the author and publisher cannot assume responsibility for the validity of all materials or the consequences of their use. The authors and publishers have attempted to trace the copyright holders of all material reproduced in this publication and apologize to copyright holders if permission to publish in this form has not been obtained. If any copyright material has not been acknowledged please write and let us know so we may rectify in any future reprint.

Library of Congress Cataloging-in-Publication Data

Levine, Evan S., 1979-
    Applying analytics : a practical introduction / E. S. Levine.
        pages cm
    Includes bibliographical references and index.
    ISBN 978-1-4665-5718-5
    1. Quantitative research. 2. Qualitative research. 3. Statistics. I. Title.

H62.L442 2013
001.4'2--dc23   2013008503

Visit the Taylor & Francis Web site at
http://www.taylorandfrancis.com

and the CRC Press Web site at
http://www.crcpress.com

"The plural of anecdote is data."
— Raymond Wolfinger

"Data is not the plural of anecdote."
— Roger Brinner

Contents

Preface   xiii

Acknowledgements   xvii

1 Introduction   **1**
   1.1   What Is Analytics?   1
   1.2   What Makes an Analyst?   2
   1.3   How Is Analytics Done?   3
         1.3.1   Deciding Which Analytic Questions to Investigate . .   4
         1.3.2   First Examination of the Data   5
         1.3.3   Reviewing Previous Work   6
         1.3.4   Applying Analytical Techniques   7
         1.3.5   Presenting Your Findings   7
   1.4   Analytics in Groups   7
         1.4.1   Collaboration   8
         1.4.2   Admitting Mistakes   9
   1.5   Onward!   9
   1.6   Exercises   11

2 Elements   **13**
   2.1   Introduction   13
   2.2   Stevens' Levels of Measurement   14
         2.2.1   Categorical Scales   14
         2.2.2   Ordinal Scales   16
         2.2.3   Interval Scales   17
         2.2.4   Ratio Scales   18
         2.2.5   Commentary   19
   2.3   Continuous and Discrete Scales   20
   2.4   Attributes   21
         2.4.1   Natural Attributes   22
         2.4.2   Constructed Attributes   23
         2.4.3   Proxy Attributes   25
         2.4.4   Commentary   25
   2.5   Data Sets   26
   2.6   Exercises   26

3 Lists   **29**

   3.1   Introduction 29

   3.2   Two Example Lists 30

        3.2.1   Heights of People in a Classroom 30

        3.2.2   Ages of U.S. Citizens 31

   3.3   Elementary Analyses 32

        3.3.1   Mean 33

        3.3.2   Median 35

        3.3.3   Mode 36

   3.4   Additional Techniques 37

        3.4.1   Histogram 38

        3.4.2   Stem and Leaf Diagrams 41

        3.4.3   Variance 43

        3.4.4   Standard Deviation 45

        3.4.5   Mean Absolute Deviation 46

   3.5   The Normal Distribution and the Height List 47

   3.6   Hypothesis Testing and Frequency Distributions 49

   3.7   Significance of Results 52

   3.8   Discarding Data from Normally Distributed Data Sets 54

   3.9   Exercises 55

4 Uncertainty and Error   **57**

   4.1   Introduction 57

   4.2   Definitions, Descriptions, and Discussion 57

        4.2.1   Illegitimate Error 58

        4.2.2   Where Does Error Come From? 58

        4.2.3   Random Error 60

        4.2.4   Systematic Error 61

        4.2.5   Aleatory and Epistemic Uncertainty 62

        4.2.6   Precision and Accuracy 63

   4.3   Quantification of Uncertainty 63

        4.3.1   Uncertainty Distributions 64

        4.3.2   Error Bars 64

   4.4   Propagation of Uncertainty 65

        4.4.1   Worst-Case Errors 66

        4.4.2   Random and Independent Errors 66

        4.4.3   Other Common Situations 67

        4.4.4   The General Case 67

        4.4.5   Non-Uniform Uncertainty in Measurements 67

   4.5   Uncertainty in the Mean 68

   4.6   Why Is This So Unsettling? 71

   4.7   Exercises 73

5 1-Dimensional Data Sets **75**

5.1 Introduction 75
5.2 Element Spacing 76
    5.2.1 Evenly Spaced Elements 76
    5.2.2 Evenly Spaced Elements with Gaps 76
    5.2.3 Unevenly Spaced Elements 78
    5.2.4 Circular Data Sets 78
5.3 Three Example 1-Dimensional Data Sets 79
    5.3.1 Voltage Measured by a Sensor 79
    5.3.2 Elevation Data 80
    5.3.3 Temperature Data 82
5.4 Interpolation 82
5.5 Smoothing 85
    5.5.1 Running Mean 85
    5.5.2 Median Smoothing 88
5.6 Fitting 1-Dimensional Data Sets 88
    5.6.1 Straight-Line Fitting 92
    5.6.2 General Linear Fitting (Advanced Topic) 95
5.7 Increasing Local Contrast 97
5.8 Frequency Analysis 99
    5.8.1 Fourier Algorithms 102
        5.8.1.1 Fourier Transforms 104
        5.8.1.2 Power Spectrum 110
    5.8.2 Lomb Periodogram 111
5.9 Filtering 114
5.10 Wavelet Analysis (Advanced Topic) 115
    5.10.1 Choosing a Mother Wavelet 117
    5.10.2 The Wavelet Transform 119
    5.10.3 Analyzing the Outputs of a Wavelet Transform 120
5.11 Exercises 122

6 Related Lists and 1-Dimensional Data Sets **125**

6.1 Introduction 125
6.2 Jointly Measured Data Sets 126
6.3 Example 1-Dimensional Linked Data Sets 127
    6.3.1 Weights of People in a Classroom 127
    6.3.2 A Second Elevation Data Set 127
    6.3.3 A Second Temperature Data Set 128
6.4 Correlation 129
    6.4.1 Scatter Plots 131
    6.4.2 Product-Moment Correlation Coefficients 134
    6.4.3 Rank Correlation Coefficients 137
6.5 Ratios 139
6.6 Fitting Related Lists and 1-Dimensional Data Sets 142
6.7 Correlation Functions 143

6.8   Exercises   148

7 2-Dimensional Data Sets   **149**
7.1   Introduction   149
7.2   Example 2-Dimensional Data Sets   151
   7.2.1   Elevation Map   151
   7.2.2   Greyscale Image   151
   7.2.3   Temperature Anomaly Global Maps   152
7.3   Restructuring Data Sets   153
7.4   Analogs of 1-Dimensional Techniques   154
   7.4.1   Interpolation in Two Dimensions   154
   7.4.2   Contour Plots   156
   7.4.3   Smoothing in Two Dimensions   159
   7.4.4   Frequency Analysis in Two Dimensions   160
   7.4.5   Wavelet Analysis in Two Dimensions (Advanced Topic)   164
7.5   Analogs of 1-Dimensional Techniques for Related 2-Dimensional
     Data Sets   166
   7.5.1   Correlation Coefficients, Scatter Plots, and 2-Dimensional
       Data Sets   166
   7.5.2   Ratios of 2-Dimensional Data Sets   169
   7.5.3   2-Dimensional Cross- and Autocorrelation Functions .   170
7.6   Higher-Dimensional Data Sets   170
7.7   Exercises   172

8 Unstructured Data Sets   **173**
8.1   Introduction   173
8.2   Extracting Structure from Unstructured Data   174
8.3   Text Mining   176
8.4   Bayesian Techniques   178
   8.4.1   Bayes' Rule   178
   8.4.2   Bayesian Belief Networks   181
   8.4.3   Bayesian Statistics   185
8.5   Exercises   186

9 Prescriptive Decision Analysis   **187**
9.1   Introduction   187
9.2   Understanding the Decision Environment   189
9.3   Determining and Structuring Objectives   192
9.4   Identifying Alternatives   197
9.5   Utility Functions   200
9.6   Decision Trees   206
9.7   Game Theory   209
9.8   Negotiations   212
9.9   Exercises   215

10 Project Management   217
  10.1 Introduction 217
  10.2 How Analytic Projects Are Typically Managed 218
  10.3 A Methodology for Project Management 219
      10.3.1 Define the Goal 220
      10.3.2 Delineate Roles 221
      10.3.3 Determine the Tasks to Achieve the Goal 222
      10.3.4 Establish the Dependencies between Tasks 225
      10.3.5 Assign Resources to Tasks and Establish a Schedule . 228
      10.3.6 Execute and Reevaluate 231
  10.4 Hastening Project Completion 231
  10.5 Project Management of Analytic Projects 232
  10.6 Other Project Management Methodologies 235
  10.7 Exercises 235

11 Communicating Analytic Findings   237
  11.1 Introduction 237
  11.2 Fundamental Principles of Analytic Communication 238
      11.2.1 Clarity 239
      11.2.2 Transparency 240
      11.2.3 Integrity 241
      11.2.4 Humility 242
  11.3 Medium 242
      11.3.1 Printed Matter 243
      11.3.2 Oral Presentations 244
      11.3.3 Elevator Speeches 246
      11.3.4 Web Pages 246
  11.4 Audience 248
      11.4.1 The Technical Specialists 249
      11.4.2 The Technical Generalists 249
      11.4.3 The General Public 250
  11.5 Tips for Data Visualization 251
  11.6 Exercises 254

12 What to Do Next   255

Bibliography   259

Index   265

Preface

I first considered writing this book when I was driving across the United States, from Berkeley to D.C. I had just finished my Ph.D. in astrophysics and was moving East to take a job working for the federal government. I spent a lot of time on that drive thinking about what exactly I had learned in graduate school and wondering why it took four years to finish my dissertation after I had completed my coursework.

One of the questions I asked myself was how long it would take to finish the same amount of work if I had been an experienced analytics professional from the start. In other words, if I had begun my dissertation with all of the analytical skills and experience required, how long would it have taken to actually complete the work? How much additional time did I invest in training myself to become an analyst? I suppose that if I had already been an analytics professional completing my dissertation it would still have taken me around half the time; the remaining two years were required to actually master the necessary analytical skills, a process that sometimes felt like banging my head against a wall.

Like most graduate students, I began analyzing data without any training beyond a few formal mathematics and statistics courses. I learned how to be an analyst the way that most people do nowadays: I was turned loose on a bunch of data and experience was my primary teacher. Though effective, learning analytics from experience is inefficient, likely to lead to mistakes, and potentially full of gaps. If a group of analysts each learn from experience on unique data sets, it's very difficult for our profession to develop a common set of techniques and practices. It doesn't have to be this way; we can do a much better job of communicating our lessons learned and establishing a common foundation.

As a step towards this goal, this book is intended to be a guide for the beginning quantitative analyst and is focused on communicating practical experience. Unlike many other books in this field, I will not provide the derivations or the mathematical justifications for the analytic techniques I discuss. Instead I will concentrate on the interpretation of the techniques, the challenges you may encounter in their use, and their strengths and weaknesses. Wherever possible, I will direct you to other references where you can study the mathematical derivations. Many authors argue that in order to properly use a technique, you must understand how it is derived. I agree with the sentiment behind this statement; if you don't understand the derivation behind an equation, you are likely to misuse it or fail to recognize an occasion where you are being misled by some feature of the data. However, there is no lack of textbooks where you can learn these derivations – it's the practical experience that's hard to find written down. Starting with worked examples is also the quickest route to applying analytics yourself.

Analytics is a growing field; more and more people are being asked to interpret data and identify understandable trends. It's no longer limited to quantitative fields such as physics or finance. For example, this book is appropriate for:

- An aspiring scientist encountering real data for the first time,

- A businessperson wanting to feel more comfortable analyzing data,

- A decision maker wanting to become an informed consumer of analysis,

- A practicing financial analyst dealing with data every day,

- A marketing expert wanting to review the basics and learn new tricks,

- A web developer looking to understand trends in page hits, and

- An experienced analyst looking for new ideas and a different perspective.

This book can be used in an academic setting or by readers working on their own to learn new techniques. In the academic setting, this book should be used as a supplement to a more formal textbook in a junior, senior, or graduate level data analysis or statistics course. For analysts reading on their own, welcome! I've cut out the boring stuff and kept the material you really need to know. For both settings, a rudimentary knowledge of calculus will be extremely helpful in understanding the topics covered, and a few select topics rely on basic linear algebra. Topics involving more sophisticated mathematics are labeled "advanced."

Beginning analysts, regardless of their field of study, benefit from practical lessons in approaching real data. For this reason, I have chosen examples that don't require specialized background knowledge to understand. By studying examples, you'll understand the circumstances in which to apply each technique, and where they can hit snags. Developing a library of reliable, tested techniques is also an important step in becoming an analyst. Solved examples will help you apply and understand your own analytic routines so you will be prepared to use them on your own data sets. You can find links to the data sets used in this book at http://www.applyinganalytics.com.

These days, most data analysis takes place on a computer, but you don't need to know how to write your computer programs to benefit from this book. However, even if you aren't interested in learning how to write computer code, you still need to know how the techniques work so you can recognize situations in which they will fail. I understand that many readers may only have Excel available for use – you can get by with it for the first few chapters, but it will start to become annoying when analyzing multiple 1-dimensional data sets. I've also found Excel's built-in plotting routines to be very difficult to fine tune, so a more flexible piece of software for visualizing data (such as Tableau) would be helpful.

If you do know how to program, this book is agnostic to your choice of computer language, in that the small amount of pseudocode isn't written in a particular language. Personally, I use Interactive Data Language (IDL) to do my own analyses, but that choice of language stems mostly from my background in astrophysics, where IDL is commonly used. MATLAB^{TM}, Python, R, S, SPSS, and C++ are also appropriate choices for the data sets in this book. You'll want to make sure you have the plotting packages installed for whatever language you choose, as much of this book is focused on best practices in visualizing data.

Though I do prefer if you code the techniques yourself, many of the languages above have built-in routines that will apply the techniques discussed in this book. If not built-in to the languages themselves, they are often available in public libraries. Python in particular has a very robust set of freely available libraries at http://pypi.python.org/pypi/, but the free Enthought Python distribution also has most of what you would need. For C++, the GNU Scientific Library is a good place to start.

When analyzing large data sets beyond the ones used as examples in this book, you'll eventually want to download and use the built-in or library techniques because they will be more computationally efficient than the versions discussed herein. However, if you know how to program I still think it is a worthwhile exercise to code your own versions so you can learn how the techniques work.

Beyond the marketable technical skills studying analytics will give you, I hope you will come to realize that analytics is actually fun. There aren't many ways in this world for someone to be an explorer without leaving their desk, and becoming an analyst is one of them. I hope this book helps to get you started on a long and fulfilling career using analytics!

Acknowledgments

I would like to thank several people for their help in bringing this manuscript to print. Julie Comerford, Barry Ezell, Derrick Nelson, Karin Sandstrom, Conor Laver, Katie Peek, and Jen Small gave me feedback on early drafts of this book. Their comments improved the writing and content tremendously. Additionally, throughout my education and career I have been very lucky to learn from some terrific teachers, among them Ralph Keeney, Daryl Morgeson, Leo Blitz, Carl Heiles, Martin White, Jim Moran, and John Wells. Their lessons and experience have had a strong influence on my thinking about analytics.

At Taylor & Francis, I am grateful for the assistance and guidance of Bob Stern, Bob Ross, Marsha Hecht, Samantha White, Kevin Craig, Shashi Kumar, and Kat Everett. The manuscript was written in LaTeX using BibDesk and Richard Koch's TeXShop. For LaTeX help, I relied heavily on my well-worn copy of Kopka and Daly (1999).

When I moved from academia to practicing analytics, I had the good fortune of starting at the Department of Homeland Security's Office of Risk Management and Analysis. The environment of creativity and energy I found there is something I've been trying to recreate since I left. I would particularly like to thank Tina Gabbrielli, Scott Breor, Robyn Assaf, Steve Bennett, Julie Waters, Tony Cheesebrough, Debra Elkins, Bob Kolasky, Allen Hickox, Sarah Ellis Peed, Ashleigh Sanders, Allen Miller, Sandy Ford Page, Natasha Hawkins, Dave Reed, Bob Hanson, Erin Gray, Matt Stuckey, Charles Rath, Lilly Gilmour, and Ryan Williams. At the New York City Police Department's Counterterrorism Bureau, I have the privilege of working with a group of consummate professionals who occasionally take a break from securing the City to listen to me ramble about the difference between variances and standard deviations. I would like to thank Richard Daddario, James Waters, Richard Falkenrath, Jessica Tisch, Rich Schroeder, Andrew Savino, Salvatore DiPace, Mark Teitler, Courtney Mitchell, TJ Brennan, Aviva Feuerstein, Greg Schwartz, Nat Young, and Ryan Merola.

Finally, this book would not have been possible without the love, support, and patience of Nicole, Ian, and my parents.

1

Introduction

1.1 What Is Analytics?

Analytics is the reduction of data to understandable findings. This is an intentionally broad definition; the data aren't required to be numerical, computers and code aren't necessarily involved, and the subject of the data isn't restricted to any particular fields.[1]

A team from IBM (Lustig et al. 2010) proposed a classification scheme for types of analytics that has been widely adopted and popularized; this scheme is useful when explaining to people without technical expertise what topics are included in analytics. The three categories are:

- Descriptive analytics,

- Predictive analytics, and

- Prescriptive analytics.

Predictive and prescriptive analytics together are also known as advanced analytics. These categories are most appropriate when used to characterize different types of analytics in a business setting, but they can also be applied to analytic efforts in other fields, such as the sciences. Let's discuss each category in turn.

Descriptive analytics refers to the use of statistics and other techniques to understand data from the past. Descriptive findings focus on what has happened, how something happens, or what is happening right now. A commonly occurring application of descriptive analytics is the use of statistics on a set of measurements. For example, a ranking of baseball teams by their winning percentage is descriptive analytics.

Predictive analytics goes a step further by extrapolating trends to produce findings regarding what will happen next or what is hidden in data that have not yet been analyzed. Forecasting future trends in the global climate, based on current data combined with simulations and models, is an example of predictive analytics.

---

[1] INFORMS, the preeminent professional society for operations research and management science, has an analytics section that defines analytics with a similarly wide scope: "the scientific process of transforming data into insights for making better decisions."

Prescriptive analytics takes the final step and determines (or, more loosely, informs) what actions should be taken based on the data. Prescriptive analytics is concerned with choosing the best course of action from a set of alternatives. This type of analytics is frequently used in the business world, but it also appears in scientific fields when connected to policy. Consider, for example, an analysis of global climate change prepared for a political leader containing information on the tradeoffs involved in adopting various versions of a carbon dioxide emission reduction treaty.

There is no reason to constrain an analysis to just one category; it's common for analytic applications to combine aspects of two or more of these categories. Additionally, though it is certainly common to begin with descriptive analytics (see SAS (2008) and Davenport and Harris (2007) for more detail), there isn't necessarily a progression from one type of analytics to another.

As an example of a project combining all three types of analytics, consider an analyst preparing a report for a customer who is deciding whether or not to purchase stock in a company. The analyst could begin by compiling information and analysis on the company's past behavior, its current stock price, and whether it typically exceeds or fails to meet its own earnings forecasts. These are all retrospective analyses that are part of descriptive analytics. Next, the analyst could try to predict how the stock will behave in the future – are anticipated market conditions similar to situations when the stock increased in value in the past? Are current market trends likely to create a more favorable environment for the company? These are both examples of predictive analytics. Finally, the analyst could determine the degree to which the forecast he assembled for the company's stock matches the customer's investing goals, summarizing this analysis with a recommendation to the customer of whether to buy or stay away. This last step is a foray into prescriptive analytics.

1.2 What Makes an Analyst?

Who actually does analytics? In my view, a broad definition of analytics must be coupled with a similarly broad definition of an analyst. When I use the term "analyst," I'm not just referring to people who spend all of their time looking at data and writing computer code. Everyone who interprets data, even if it's only on rare occasions, is an analyst. The broad definition of analytics also requires that analysts possess a wide variety of skills; an analyst can't be limited to just digging around in data and applying numerical techniques. In other words, an analyst does more than interpret data for people who are not fluent in numbers.

A good analyst should also be able to communicate the import and the origin of findings. A conversation with an analyst should not be like consulting the oracle at Delphi, where mysterious truth is handed down from above; after the discussion of the data is finished, the people listening should be able to explain the work nearly as well as the analyst who actually performed the work.

A good analyst is also a good manager, both in terms of managing people and managing projects. Successful analysts develop managerial skills to increase their effectiveness and efficiency, while unsuccessful ones neglect these skills and instead do everything on their own. They then wonder how everyone else is so much more productive; I know this, because I've been there myself. Fortunately, management skills can be learned, and a little bit goes a long way.

However, being an analyst certainly requires technical skills in addition to communicative and managerial skills. Luckily, even though no two data sets are identical, to analyze a data set you generally don't need to invent a unique set of analytic tools. Many techniques are applicable across a variety of types of data sets, regardless of the details of the data. As important as using these techniques to identify findings is knowing what your data can't tell you. This will prevent you from overreaching in your analysis and being tricked into believing that a random fluctuation in the data represents something profound.

The proliferation of data has made this portfolio of analytical skills widely relevant to many aspects of scientific, business, and governmental work, and thus increasingly in demand. Academic and professional institutions are now beginning to adapt to the new environment. For example, the professional society INFORMS recently began holding an annual conference on business analytics, and has just started a certification program for quantitative analysts. In the United States, the McKinsey Global Institute forecast a shortfall of 140,000 to 190,000 highly qualified analysts by 2018, plus an additional shortage of 1,500,000 managers with the skills to use the findings analysts produce. Other evidence of the growth in analytics includes recent influential books, such as Tom Davenport and Jeanne Harris's *Competing on Analytics*, which highlights the increasing importance of analytics in the business world, and the popularization of modeling for elections and sports, exemplified by Nate Silver's articles in the *New York Times* and his book *The Signal and the Noise*.

1.3   How Is Analytics Done?

Before jumping into the details of the analytical techniques, it's worth taking a brief moment to discuss the process of performing analysis. There is much more involved in completing an analytic project than just the technical work. Here, I will briefly describe how to decide which analytical questions are worth pursuing, how to start analyzing data, the review of previous work, the application of analytic techniques, and the presentation of findings. Analytics projects don't generally proceed through these steps linearly, but instead tend to jump back and forth from one step to another. For example, an analyst might present work to a supervisor several times throughout the course of the project, then return to applying analytical techniques to incorporate feedback. Still, in a complete analytics project, each of these steps will at least be touched upon.

1.3.1   Deciding Which Analytic Questions to Investigate

Most analysts, with the possible exception of those just beginning in their respective fields, have some choice regarding the questions they study and the projects on which they work. There is a skill to making these decisions – without a doubt, making wise choices regarding which problems to study makes a big difference in an analyst's career.

Attractive questions and projects possess at least one (and hopefully, several) of the following characteristics:

- They have high potential impact.

- The data necessary are readily available or already in hand.

- They are of high importance to your organization's leadership (or they pertain to a "hot" topic in a scientific field).

- There is relatively little time and/or resources required to get a result.

- For academics, there are few competitors working on the same question, minimizing the chance you'll be scooped.

- There is a relationship to an earlier project you've worked on that gives you a competitive edge over other analysts.

- Completing this project will give you a competitive edge in future attractive projects.

- For whatever reason, the question is of high intellectual interest to you.

Unattractive analytic questions and projects, unsurprisingly, have the opposite qualities.

Don't feel as if you must answer every analytic question that comes your way or that you need to follow-up on every idea you have. Your analytic effort is a scarce resource and you need to optimize where you invest your time. For example, it can be an excellent decision to decline a project that would have a small impact on your organization if it will take a lot of time to complete, even if it is certain to provide a definitive result. The most important thing is for you to understand why or why not you've chosen to work on a particular project and to make these decisions in a rational way.

In practice, most analysts work on several projects simultaneously. This means that an analyst can put together a portfolio of projects that mixes high-risk high-reward projects with low-risk low-reward projects. Note that low-risk high-reward projects actually exist; you just have to find them! Similarly, high-risk low-reward projects should be avoided.

Once you have data in hand, the range of questions you can investigate gets considerably narrower. How do you decide what questions to ask of the data? Many analysts, when they first look at a data set, ask themselves: "What are the data trying to tell me?" This is exactly the wrong approach. The data are not trying to tell you anything. Many data sets are of poor precision, improperly collected, and more dangerous than they are useful. Sometimes the right answer is to realize you can't learn anything from a data set, and just toss it out.

To put yourself in a better frame of mind, rephrase the way you ask your analytical questions. A more appropriate question to ask is: "What can I learn from this data set?" Or, even better: "What decision am I trying to make based on these data, and how should this new information influence my decision?" This last question, an example of prescriptive analytics, is the purview of the field of decision analysis, and is particularly relevant for analysts operating in the business world. However, it is also useful for thinking analytically about your own life choices and for managing your career as an analyst. I provide a brief introduction to prescriptive decision analysis in Chapter 9.

1.3.2   First Examination of the Data

As a beginning analyst, when I first got my hands on a data set, I felt two things: despair at how complicated the data looked and overwhelmed by how much work needed to be done. I suspect these feelings are more or less universal, but they lead to unproductive behavior – the application of whichever analytical techniques are available to the data without rhyme or reason. This is the wrong way to do analysis, but it's how many analysts operate.

The first and most important step when analyzing a data set is to examine the raw data thoroughly. Take a deep breath. Make simple plots or maps of it, display it backwards, upside-down, and sideways, get to know the major features. Plot it on linear and logarithmic axes. One way to test whether you have sufficiently studied your data is to try and draw the basic structure freehand, without actually looking at the data. You should be able to draw a fairly accurate map from memory before proceeding.

Identify "special" areas in the data. Are there edges? Places where the uncertainty is higher or lower than usual? Regions where strong discontinuities or jumps are prevalent? Are there many outliers in the data? Why or why not? Make notes of these features and return to them when you are stuck on a problem. Could one of these features be causing it? When I was a graduate student, Prof. Hy Spinrad made a point of always asking about the data point that stood out the most; there is often more to be learned from what doesn't fit than from what does.

There are several reasons to study the data set in this way, before applying any analytical techniques. First, when it's all said and done, you're going to present your findings to your boss, your advisor, or a larger audience. In these settings, you'll need to be able to respond to questions about the data without having access to your computer or your notes. You'll even be asked to speculate on the fly about techniques you haven't actually applied to the data! The only way to answer these questions with confidence is to know your data backwards and forwards. Thoroughly knowing your data also helps when looking for mistakes in your analysis; the special areas in the data you've identified are the most likely regions to cause problems with the algorithms you apply.

1.3.3   Reviewing Previous Work

Next comes a step that is absolutely necessary, but somewhat of a drag: searching the literature, the Internet, and other reference texts for similar and related work that has been done in the past. Doing this well will save you lots of time over the lifetime of the project, especially if you're not interested in inventing a new solution, but rather to just find any solution that works. Take note of the techniques with which previous analysts have approached this problem, as well as their results. What sorts of methods have yet to be applied in this area? If possible, discuss the potential project with an expert that you trust.

For those analysts who must be innovative, e.g., those that must publish in academic journals for career advancement, performing the literature search can be another source of dread. When I am doing academic research, I worry that the next journal article I read will contain the idea on which I am working, meaning that my research is not something new. That's not a reason to skip the literature review. It's still important to search the literature thoroughly because if someone reviewing your paper is aware of previous work that you didn't find, you will have wasted a lot of time when your paper is rejected for being duplicative.

Some people argue that learning about the work that has already been done on a problem reduces the chance of your coming up with a creative new idea. This argument has some merit; one way to minimize its effect is to brainstorm ideas before starting the literature search. Return to your list of ideas after completing the search and determine which ideas are novel.

For problems for which a lot of previous work has been done, it can be hard to know when to stop the literature search. In these cases, it isn't realistic to try and read everything that has been written on a subject. Wilson (1952) describes two essential criteria for ending your search: establishing a general baseline of knowledge in the field and ascertaining definitively that an identical or very similar project hasn't previously been completed.

1.3.4 Applying Analytical Techniques

Only after you've performed all of these preliminary steps in the process is it time to apply analytical techniques. The majority of this textbook is a discussion of these techniques. Start by applying the simplest, such as mean and median, and then proceed to more sophisticated techniques only if necessary. Using complicated methods to identify findings that could have been produced more straightforwardly is a surefire way to introduce a bug into your analysis and get something wrong.

Often the process of performing analysis can get complicated; for example, you may have to test, build, and apply several simple analyses before you make a decision regarding which more sophisticated analytical technique to pursue. In other words, the different tasks that make up a project don't always follow linearly one after another. For this reason, every analyst should have a basic understanding of project management, discussed in Chapter 10. These techniques are actually not commonly used in analytics; your adoption of them will be another skill to make you more efficient than the old guard.

When your analysis nears completion, return to the list of special areas you identified in your thorough examination of the raw data. Have you taken all of these special areas into account? Can you now explain how they all originated? If not, why not?

1.3.5 Presenting Your Findings

When your analysis is done, you will stand up in front of an audience and present your findings, write a research paper on the data, or brief an executive on your findings. At this point, you must know your data better than anyone in your audience. Furthermore, you will need to *know* that you know it better; since the vast majority of your audience will never try to confirm your findings, the confidence you project in your own work will be an important factor in whether they accept your conclusions. Tips on how to best communicate your analysis are found in Chapter 11.

1.4 Analytics in Groups

Luckily, when making their way through these steps in the process of analytics, analysts are rarely on their own. Analysts working in collaborations can accomplish much more than those working independently, but being part of a team means that each analyst has a responsibility to the group when it comes to admitting any mistakes they might make. Though these topics are not formal steps in the process of analytics, knowing how to approach them makes any project proceed more smoothly.

1.4.1   Collaboration

In modern analytics, many projects are carried out by groups with the tasks split up amongst collaborators. It's a cliche, but communications technologies are a key ingredient in these collaborations. Can you imagine coordinating amongst a team of analysts without email and file sharing? With the speed at which these technologies have developed since the 1990s, it's no wonder that we're still working the kinks out of analytical collaborations.

There are numerous benefits of effective collaboration. Analysts can choose which tasks they perform according to their interests and skills, leading to increased capacity and efficiency. Analysts generate ideas from talking to each other, so having multiple people look at the same data often leads to new and creative approaches.

Well, that's the idea, anyway. In reality, there are added complications that eat into these productivity gains: work has to be coordinated, no single individual is an "expert" on the project, and tasks can fall into the gaps between people's responsibilities and not get done. Most importantly, the unpleasant, boring, and repetitive tasks also need to get divided up in a fair way. Self-sacrificing leaders will sometimes take all of these unpleasant tasks upon themselves and build up resentment towards the rest of the group, or uncaring managers can feel that these unpleasant tasks are not their responsibility and delegate them all downwards, leading to resentment amongst the group members. Finding a happy medium is key.

Good management is often the difference as to whether the benefits of collaboration outweigh the drag of the complications. I've supported many projects throughout my career, and often can tell within 5 minutes how the project is going to fare from the organizational and management skills of the project lead. I would expect that you've had similar experiences working on group projects with fellow students or coworkers. When you participate in these projects, take notes as to what works well and what doesn't. Emulate the behaviors that are successful when you run your own projects.

When offered the chance to participate in collaborative projects, I recommend volunteering for at least one task that is outside of your established skill set. This will give you an opportunity to learn a new technique and receive prompt and constructive criticism on it from your collaborators. After all, it's in their interest for you to do a thorough job and make them look good! Even better are the opportunities to learn a skill when there is someone else in the group who is an expert in that skill; you can pick up tips and tricks from that person much faster than you would learn them on your own. Continuing to add to your skills is also very important; it's not good for your career to be pigeonholed as a narrowly skilled analyst.

1.4.2 Admitting Mistakes

Being an analyst isn't all fun and games. At times, you're going to make mistakes, even if you are a top-notch analyst. It's inevitable. Computer code is complicated, analytical techniques are subtle, and sometime randomness enters into data in ways that seem diabolical. Speaking personally, I've made plenty of mistakes over the course of my career, some of them quite embarrassing.

My advice is to admit your mistake to your supervisor and team as soon as you have confirmed it occurred. First, summarize what went wrong along with its implications. Then describe in detail what caused the mistake to happen, how the mistake connects to the analytical outputs, and what influence it has on your final findings. Depending on the personality of the person you're presenting to and the impact of the mistake, this kind of presentation can be an unpleasant experience.

It's best to present a plan for correcting the error at the same time as you explain how it happened. Pointing out an error and saying you don't know how to fix it will likely lead to your superiors and colleagues losing confidence in you. Make sure you mention how much time it will take you to correct the mistake and what additional resources, if any, will be required to fix it. It's also good practice to discuss how you can prevent mistakes of this type from happening again.

So, you've got to take your medicine when you make a mistake, but that doesn't mean you should allow yourself to be treated poorly when you admit to one. As I've said, even the best analysts make mistakes, and your supervisor and colleagues should know that. If you find yourself in a situation where you are seriously worried about how someone is going to respond to your admission, it may be time to look for a new place to work. After all, analytical skills are in demand and you deserve to be treated well for making good faith efforts. Life is too short, and analytics too much fun, to settle for anything less.

1.5 Onward!

The majority of this book, beginning with the next chapter, is occupied with the discussion of technical analytic skills.

Chapter 2 defines the element, the basic unit of analysis I will use in this book. I then present three categorizations that are helpful in making sense of the variety of data that analysts encounter.

Chapter 3 explores lists, a simple type of data structure in which the order of the elements does not matter. The interpretation of elementary analytic techniques (mean, median, and mode) is covered in depth, followed by more advanced techniques. A series of topics related to the normal distribution closes the chapter.

Chapter 4 is an introduction to error and uncertainty in analytics. This topic is fraught with inconsistent definitions, so the chapter begins with a discussion of the relevant concepts. Next, I cover the mathematics of error bars and the propagation of error, which are foundational concepts for the communication of uncertainty. The chapter ends with a warning regarding the unexpected breakdown of models.

Chapter 5 returns to the discussion of data structures, focusing on 1-dimensional data sets. These structures occur with a variety of element spacings, so the chapter begins with a categorization of those spacings. Interpolation, smoothing, and fitting are then applied to example data sets. I then introduce frequency analysis, an important tool for analyzing 1 and higher dimensional data sets. After a brief introduction to filtering, the advanced topic of wavelet analysis, which combines frequency and positional information, closes the chapter.

Chapter 6 covers techniques appropriate to the analysis of related lists and 1-dimensional data sets. Analysts often have two (or more) related data sets they must compare, and there are some specialized techniques available for that purpose. Correlation, ratios, fitting, and correlation functions are all discussed, with examples.

Chapter 7 focuses on 2-dimensional data sets, the simplest multidimensional data structure. The presentation in this chapter proceeds in parallel to the previous two chapters, discussing the generalization of techniques that were previously applied to 1-dimensional data sets and related 1-dimensional data sets. The chapter ends with examples of data sets with more complicated dimensionality.

Chapter 8 discusses unstructured data sets, which often require specialized techniques to analyze. I first present the basic strategy for extracting structure from unstructured data, then I use text mining as an example of unstructured data analysis. Next, I present Bayesian techniques as an alternative means of analyzing unstructured data and conclude the chapter with a brief discussion of Bayesian statistics.

Chapter 9 pertains to the use of analytics to inform decisions, a field known as prescriptive decision analysis. Structured techniques are available to inform decisions; I give an in-depth presentation of the classic technique of multicriteria decision analysis and the related technique of decision trees. I then provide brief introductions to two alternative approaches to decision analysis: game theory and negotiation analysis.

The following two chapters of the book are dedicated to the other skills you will need as a professional analyst. This material is just as essential to your success as the technical subject matter pertaining to data elements and structures.

Chapter 10 provides an introduction to project management, which is essentially the application of analytics to the process of doing work. Examples of applying project management to several types of projects are discussed.

Chapter 11 is centered around the communication of analytic ideas. I first present some guiding principles for analytic communication, then in broad strokes I categorize the various media that analytic communication occurs over and the audiences with which analysts communicate. The chapter closes with some helpful tips for data visualization, a common task analysts must perform.

Chapter 12 concludes the book with suggested methods for continuing to expand your analytical skills.

1.6 Exercises

1. Have you ever used analytics to inform a personal decision? If so, describe that decision and the analysis you carried out. If not, what personal decisions do you think analytics would have been helpful for?

2. Describe a situation in which you've used each of the three types of analytics.

3. Briefly describe a group project in which you have previously participated. What were some positive aspects of how the project was run? What were some negative aspects? What would you do differently if you had to do it over again?

4. Briefly describe a mistake you made at a job or on a project. How did you handle that mistake? With the benefit of experience, how would you handle it differently?

2

Elements

2.1 Introduction

The world is filled with myriad pieces of data. From the color of your car to the number of dollars in your bank account, data are everywhere. Let's call each piece of data an <u>element</u>. A single element, by itself, is not very interesting because there isn't anything to compare it to, but elements are the building blocks for the more complex data structures I will discuss later on in this book. When a group of elements is arranged into a structure, the elements together can be called a data set.

Elements contain data or, alternatively, information. However, there are constraints on the type of data that a particular element can hold. For example, it wouldn't make any sense for an element associated with a temperature measurement to contain the value "green." Similarly, an element associated with the number of soccer balls I own couldn't contain the value 2.52. The range of possible values that a given element could contain is known as its <u>scale</u>. Scales often have an associated physical unit, such as meters or degrees.

The data stored in an element and the scale with which it is measured influence how that element can be analyzed. For example, you can't divide green by blue and get a result that makes sense, but you can divide the balance in your bank account by 10 without the math being an issue (you may have other objections to that transaction). <u>Classification schemes</u>, which group elements by common characteristics, can explain why certain elementary mathematical operations make sense for some elements but not for others.

There are innumerable different classification schemes for elements. This chapter will discuss three separate classification schemes that capture different aspects of element behavior. The schemes I have chosen are helpful for characterizing elements, but no scheme divides the spectrum of elements into categories without some grey areas; there are just too many varieties of elements to reduce them to a definitive handful of types. Not only that, but there are multiple sets of characteristics by which it is helpful to classify elements. Adding to the confusion, many schemes have inconsistent definitions across the literature.

Because of the lack of basic data analysis classes in most undergraduate curricula, many beginning analysts do not have a firm grounding in element classification schemes. This is a shame, because even a quick overview will positively influence the design of your analyses. Experienced analysts will have an intuitive understanding of parts of this section, but examining different types of elements in a structured way can provide ideas for more analyses and a better intuition for the type of information contained in the data. Also, you'll be surprised by how often thinking about element classification schemes can clarify problems in your analysis. People often fail to understand the elements they are analyzing and perform misleading and incorrect analyses as a result.

Let's start by discussing my favorite classification scheme, one that has fallen out of favor in recent years but still yields valuable insight.

2.2   Stevens' Levels of Measurement

Stevens (1946, 1951) developed the element classification scheme known as levels of measurement in the midst of a lengthy debate in the British scientific community on the nature of measurement. Stevens' scheme classifies elements according to the characteristics of the scale with which they are measured, placing the scales into one of four groups: categorical, ordinal, interval, and ratio. There are certainly elements that don't fall neatly into one category or another, so don't treat this as a hard and fast classification. I'll proceed by discussing the most mathematically limited scale first and then move on to scales that allow more mathematically sophisticated operations.

2.2.1   Categorical Scales

Categorical scales, also known as nominal scales, are constructed from descriptors with no mathematical relationships. Examples of categorical scales you may be familiar with include shapes (square, circle, and triangle), colors (green, blue, red, yellow, black, and white), fruits (apples, pears, and bananas), astronomical objects (neutron stars, black holes, comets, and planets), and student majors (Chemistry, History, and English).

For a more detailed example of a categorical scale, imagine you were studying a genetically inherited disease, such as cystic fibrosis, in which symptoms result when a person has two copies of the disease gene. Suppose you wanted to classify each member of a population sample into one of three categories:

- Symptomatic (those who have two copies of the disease gene and suffer from the disease),

- Carrier (those who have one copy of the gene for the disease and can pass it on to their offspring but do not demonstrate symptoms), and

- Non-carrier (those who don't carry any copies of the gene).

In this example, each person in the sample is tested and classified as either symptomatic, a carrier, or a non-carrier. After the tests have been carried out, the only sensible mathematical operation is to count how many people there are in each category. There are no relationships between the category descriptors, such as "carriers are greater than non-carriers." Similarly, you cannot multiply the categories and find the answer to "symptomatic times carrier." Categorical scales don't have these relationships (note that I am talking about the categories themselves, and not the counts of the number of people in each category). And since these simple operations don't make any sense, there certainly is no way to take an average, a median, or a standard deviation of the descriptors. The best you can do is to analyze the population count of each category and draw conclusions from those data.

In a categorical scale, the categories must be mutually exclusive (see inset for discussion). For example, a person in the cystic fibrosis sample can be a carrier or symptomatic, but not both. Categorical scales should also be collectively exhaustive, meaning there should be a category for every element to be classified (see second inset for discussion). For example, if there were a person in the cystic fibrosis sample who had two copies of the gene but did not demonstrate symptoms, the "symptomatic, carrier, and non-carrier" categorical scale would not be collectively exhaustive. You could fix the scale by adding "two disease genes but asymptomatic" as a category or by introducing an "other" category. Using the "other" category has the added benefit of "failing safe," meaning that if people with yet another genetic makeup are present, it would break the categorical scale because they could be categorized as "other."

Mutually Exclusive

Two categories are mutually exclusive if no element can be placed into both categories. Similarly, the categories that make up a categorical scale are mutually exclusive if no element can be placed in two or more categories. Most of the time, it is easy to judge whether a particular categorical scale meets this criteria, but there can be situations where an element that would break mutual exclusivity can occur but happens to not be present in a given data set. In that case, the categories can be treated as mutually exclusive for that data set, but aren't mutually exclusive in all circumstances.

Collectively Exhaustive

A categorical scale is collectively exhaustive if every element can be placed into a category. Similar to mutual exclusivity, there can be situations where an element that would break the collectively exhaustive property can occur, but is not present in a particular data set. Then, the categorical scale can be treated as collectively exhaustive for that data set, but that scale can run into trouble with other data sets.

The versatility of categorical scales is what makes them so powerful. In fact, many fields where qualitative analysis dominates are using categorical scales by another name. For example, let's say you were a business intelligence analyst trying to predict whether a rival software company is going to develop a competitor to your company's most popular product. You could examine your rival's public statements, new hires, and investment strategy to place the company into one of two mutually exclusive and collectively exhaustive categories: developing a competitor or not developing a competitor. This seems too obvious to be useful, but in Section 8.4.2's discussion of Bayesian networks it will become clear how powerful this technique is. That said, if an analysis relies entirely on categorical scales, most traditional analytic techniques will be difficult to apply.

Categorical scales can also incorporate uncertainty, for example, the uncertainty resulting from mistaken categorizations that can be described with a set of probabilities. A situation in which this could occur would be a blood test for the gene that causes a genetic disease. After testing a patient, you could imagine describing your state of knowledge as follows:

> The patient is not sick, so there is no chance he is symptomatic. The test came back negative for the gene, but the test doesn't always give an accurate result. Based on my experience with negative results in the past and the prevalence of this gene in the general population, I'd say I'm 95% sure this person is a non-carrier, but there's still a 5% chance he's a carrier. There is a 0% chance he's symptomatic.

Similar situations with real data (and real medical tests) are quite common. Categorical scales can handle this uncertainty without issue.

For those of you paying close attention, Stevens' levels of measurement, which classifies scales into categorical, ordinal, interval, and ratio types, is itself a categorical scale. However, there are some scales that could arguably be classified as one category or another, breaking mutual exclusivity.

2.2.2 Ordinal Scales

Ordinal scales are built on descriptors with simple relational properties, such as "High," "Medium," and "Low." Some common ordinal scales include letter grades from "A" to "F" (without numeric grade point average equivalents) and the star rankings of hotels or restaurants. The only strictly allowed mathematical statements are relational, such as "High" is larger than "Medium," or "4 star" is better than "3 star." Addition doesn't generally work for ordinal scales, as with categorical scales, in the sense that "Low" plus "Medium" does not equal anything reasonable. Similarly, subtraction, multiplication, and division, as well as any of the more complicated mathematical operations typically don't make sense.

Notice I said that addition "doesn't generally work." What exactly does that mean? It means there aren't any absolute rules here, because there is a long and ongoing debate regarding the defensibility of using simple statistics, such as the mean and standard deviation, with ordinal scales (Stevens 1951; Velleman and Wilkinson 1993). These two operations depend on addition and division of numbers assigned to the descriptors, such as "High" being 3, "Medium" being 2, and "Low" being 1. Assignments like these are commonly accepted in social and behavioral sciences, and as Stevens himself stated "there can be invoked a kind of pragmatic sanction: in numerous instances it leads to fruitful results." Since my goal is to provide analysts with practical advice, I'm not going to argue with a simple methodology that provides useful results; I only ask that you understand what operations you are performing on your elements, and what kind of errors those operations expose you to.

Even given a lot of leeway about what can defensibly be done with ordinal scales, from my experience they are still the most frequently misused level of measurement. Often, the categories are named numerically by the analyst, which causes others to treat the categories as if they had the same properties as integers without understanding the potential limitations. In other words, a measurement that falls into "group 1" plus a measurement that falls into "group 2" is assumed to result in a measurement that falls into "group 3," whether or not the analyst intends the audience to interpret the data in that way. Similarly, a "4 star" hotel will be assumed to be twice as good as a "2 star" hotel, regardless of the intent of the analyst constructing the scale. When using an ordinal scale, my advice is to avoid labeling the groups numerically whenever possible. Call the groups "High," "Medium," and "Low," or "Red," "Yellow," and "Green," anything but 1, 2, and 3. See what you can uncover in your data first just by using relational logic without using numerical assignments. Know that if you do use numerical assignments, the means and standard deviations you calculate do not behave the same as those calculated from interval and ratio scale elements. Above all, it is highly unlikely that you'll be able to draw defensible conclusions from taking ratios of the numerical analogs of an ordinal scale.

2.2.3 Interval Scales

Interval scales have equal distances between adjacent numbers, but their zero point is arbitrary (i.e., the zero does not represent the absence of whatever quantity is being measured). Interval scales have some of the properties you would expect from "normal" numbers, but not all; for example, they generally have defensible addition and subtraction, but not multiplication or division. They are often confused with the final measurement category we will discuss, ratio scales.

Let's discuss a few examples, starting with temperature, which is a measurement of the random atomic motion in a material. The Fahrenheit and Celsius temperature scales are examples of interval scales, because objects at $0°F$ or $0°C$ still have random atomic motion. In other words, the zero of the

Fahrenheit and Celsius scales is not the total absence of random atomic motion (known as absolute zero); rather, zero is placed at an arbitrary point. For the Celsius scale, zero degrees is the freezing point of water, and for the Fahrenheit scale it is the freezing point of a specific brine mixture. Another common interval scale is the year, measured in the A.D. or C.E. systems (because time didn't start at the year zero). The year scale is a particularly strange interval scale because there was no year 0; 1 B.C. was followed by 1 A.D.[1]

To illustrate a shortcoming of interval scales, consider the Celsius temperature scale. An object at 2°C is 1°C warmer than an object at 1°C. It is sensible to say that the temperature difference between a 3°C object and a 2°C object is the same as that between a 2°C object and a 1°C object. Contrast this with ordinal scales, where the difference between "High" and "Medium" is not guaranteed to be the same as the difference between "Medium" and "Low." However, it is not correct to state that the 2°C object is twice as warm as the 1°C object because the zero of the scale is set to an arbitrary physical quantity, i.e., the freezing point of water, rather than the total absence of that quantity.[2]

It's tricky to say which mathematical operations are strictly permissible with interval scales. Quantities which rely on the difference between measurements, like standard deviations, are sensible, but quantities that rely on taking ratios between measurements are not strictly allowed. Be careful when you apply elaborate analyses to interval scale measurements; check to make sure you understand how each step in the analysis relies on the scale's construction. One good way to test if your analysis makes sense is to map to a new scale that is the same as the old scale, with the exception of a constant offset, and check if your analysis produces the same results. For example, to check an analysis that relies on the Celsius temperature scale, create a new scale called "Delsius" whose relationship to Celsius is given by the following: degrees Delsius = degrees Celsius + 100. Do you find the same results with the new scale?[3]

2.2.4   Ratio Scales

Ratio scales, also known as absolute scales, are the measurement type most like the traditional numbers with which you are familiar. These scales have all the properties of interval scales, but have the additional property that zero is really the absence of that quantity. The ratio counterpart to the Celsius and Fahrenheit scales is the Kelvin scale, because 0 K is absolute zero, the

---

[1] In fact, these calendar systems do not precisely meet the definition of interval scales because each year is not a uniform length of time. Leap years, for example, are one day longer. This is a detail I'm ignoring for simplicity's sake.

[2] You can wiggle around this by saying it is twice as warm "relative to the freezing point," but that's not very useful and subject to misinterpretation when you present your analysis.

[3] Alternatively, you could just use the Kelvin scale, which differs from Celsius by a constant offset of 273.15.

absence of random atomic motion. This property results in the ratios of these measurements having intuitive meanings: an object at a temperature of 2 K actually has twice the random motion as an object at 1 K. Money in your savings account is another example of a ratio scale; a \$200 balance is twice as big as a \$100 balance and a \$0 balance is the absence of money.

All mathematical and statistical operations can be defensibly used for ratio scales, because all the basic operations have the meanings you would expect. However, be careful that the mathematical operations you perform don't convert your ratio scale to an interval scale unintentionally. For example, if you added a constant figure to a ratio scale, you'll convert it to an interval scale and you'll lose the ability to make sensible ratio comparisons between your elements because zero will not be the absence of the quantity anymore.

Here's an example from the world of finance that illustrates how widespread the confusion between ratio and ordinal scales is and how serious the impacts can be. Leading in to the 2008 financial crisis, financial firms traded specialized bonds built by aggregating mortgages; the bonds were constructed to ostensibly reduce their risk of default but instead just obfuscated that risk. Like other bonds, the specialized mortgage bonds were graded by ratings agencies using a cryptic scale – the bonds with the smallest chance of default graded as AAA, followed by AA, A, BBB, etc. Standard & Poor's, a rating agency, argues that these grades are an ordinal scale, without a direct equivalence to a ratio-scale chance of default. However, investors studied the history of graded bonds and found that, on average, fewer than 1 in 10,000 AAA graded bonds defaulted in their first year. Similar probabilities were calculated for the other grades on the scale and were used to construct models of financial risk for these specialized bonds. During the subprime mortgage financial crisis, many of the specialized bonds defaulted at rates much higher than the historical probabilities would suggest they should. Many trading firms blamed the ratings agencies for inaccurately grading bonds, but the ratings agencies claimed that the rating scale was ordinal and not meant to be treated as equivalent to a ratio scale.[4] Though there were many other contributing effects to the subprime crisis,[5] especially the construction of these specialized bonds and their purchase and sale by people who didn't fully understand them, this miscommunication of scales was also an important contributor (Lewis 2010; Sorkin 2009).

2.2.5   Commentary

As mentioned previously, Stevens' levels of measurement are helpful categories, but there are two reasons why the classification scheme sometimes

---

[4]Many analysts believe that the ratings agencies encouraged the use of their bond ratings as equivalent to probabilities of default and only now, after the crisis, are falling back on the ordinal scale defense to avoid liability.

[5]Silver (2012) has an excellent discussion of those other causes, such as the correlation between mortgage defaults. Really, there were quite a few mistakes made.

breaks down. First, some scales fall into gaps between the categories. Second, although when Stevens first developed the scheme he made an attempt to classify mathematical operations as "permissible" for some categories and not for others, the rules prescribing which operations are permissible are not hard and fast (Velleman and Wilkinson 1993). An especially entertaining example that illustrates both of these problems can be found in Lord (1953), which describes a group of freshman football players concerned that their assigned jersey numbers (a categorical scale) are generally lower than the sophomores' numbers. The professor who assigns these numbers struggles with how to test this hypothesis, since averaging a categorical scale is not "permissible." The point here is to not to drive yourself crazy about what is and is not permissible, but instead just to consider the operations you are performing on your elements and to take care not to abuse them.

Think back over the data you have analyzed in your career. How would you categorize your elements according to Stevens' levels of measurement? Does this cause you to reconsider the operations you used to process the data? I certainly felt that way after I read Stevens' work for the first time. As you move forward, keep in mind that, as described in Section 1.3.1, sometimes you just don't have the right data to investigate the question in which you are interested. In these situations, it is especially important not to overreach by abusing the scale with which the elements were measured.

2.3   Continuous and Discrete Scales

Another helpful classification scheme is the distinction between continuous and discrete scales. Just like Stevens' levels of measurements, this scheme has grey areas, but it can give you useful insight.[6] In the most general sense, a continuous scale can take on an unrestricted set of numerical values, while a discrete scale can only take on a value from a restricted and often finite set of numbers or other descriptors.

The vast majority of scales can easily be classified as either continuous or discrete. There is a straightforward (and mostly accurate) test to differentiate between the two types of data elements: if there are an infinite number of possible values in between the values of the elements q_i and q_j (when $q_i \neq q_j$) then the data elements are continuous; otherwise they are discrete. This test cannot be used on elements measured with categorical scales, because for these scales it isn't possible to establish whether one element is between another two.

Note that it is possible for a discrete scale to have an infinite number of potential values, such as the set of all integers, or a limited set of values, such

---

[6]Using the terminology we developed earlier, you can think of a grey area as something that causes the mutual exclusivity and collective exhaustivity of a classification scheme to break down.

as the days of the week. It is also possible for a scale to be continuous for some portion of its potential range and discrete for some other portion, but this is unusual.

Physical quantities, when not rounded or limited by measurement precision, are generally continuous.[7] Some examples of continuous elements are:

- The time it takes you to run 100 m,

- The frequency of a particular audio pitch, and

- The weight of a cantaloupe.

Discrete scales are often associated with rounded numbers or counts of objects and events. Examples of elements measured with discrete scales include:

- The number of people in each country in the world,

- The birthdays of everyone in your first grade class,

- The heights of all the people at a university, rounded to the nearest inch, and

- A binary file, where each element is either a zero or a one.

In practice, nearly all elements we deal with computationally are technically discrete. Even if they are measured from real continuous physical quantities, they have been discretized by instrument precision or rounding so that the data can be entered into a computer.

Data sets measured with continuous scales are analyzed and visualized differently than those that are measured with discrete scales. The distinction drawn by this classification scheme will influence the analytical techniques applied throughout this book.

2.4 Attributes

The final element classification scheme I will introduce is drawn from the field of decision analysis (a topic I will discuss in depth in Chapter 9). In decision analysis, <u>attributes</u> are measurements that provide information about some behavior of interest. That behavior of interest could be almost anything: a person's wealth, the temperature of a star, and the quality of a house are just a few examples. The behavior itself is what the analyst is really interested in – attributes provide the window through which the behaviors can be measured.

---

[7]I'm ignoring the discretization that can result from quantum mechanics or the weight of a single atom. Like I said there are some real grey areas here, but don't miss the forest for the trees.

Attributes can be grouped into three types: natural, constructed, and proxy (Keeney 1993; Keeney and Gregory 2005). Though these attribute categories weren't intended as a general element classification scheme, I find them helpful in illustrating important concepts regarding the nature of measurement because the scheme classifies elements and their associated scales by the distance between the measurement and the behavior of interest. The simplest type of attribute is natural: one in which the measured quantity is a direct reflection of the behavior of interest. However, sometimes there isn't an obvious way to measure that behavior, either because the behavior is hard to quantify or because it can't be directly observed – then a constructed or proxy attribute must be used.[8]

2.4.1   Natural Attributes

Natural attributes directly measure the behavior of interest. Natural attributes usually have straightforward meanings and are in widespread use by analysts studying the same behavior. There can be multiple natural attributes appropriate to measure a single behavior.

Many analysts assume that all of their data are measured with natural attributes, but the connection between what you can measure and the behavior of interest isn't always direct. Furthermore, there are many behaviors that an analyst may want to measure for which no natural attribute exists or for which obtaining access to the natural attribute is prohibitively difficult.

I'll quickly describe an example of a natural attribute so I can compare it to the other types of attributes. In this discussion, I'm going to use some simple financial concepts from the stock market. Stock prices are a readily available attribute, but what type of attribute are they? That depends on the behavior of interest. If you're interested in the perceived value of a company, stock prices (coupled with a knowledge of the number of existing shares) are natural attributes because they represent the open market price of a fixed percentage of the company.[9] If, on the other hand, you were trying to measure some other sort of behavior, the exact same stock prices could be a different type of attribute, as I will demonstrate.

As an example of a behavior for which no natural attribute is readily available, consider the intensity of public support for a proposed law. For many analytic questions, such as trying to predict how much money could be raised to advertise for the law, the obvious attribute of the percentage of the public that is supportive isn't the best attribute to use, since that percentage doesn't tell you about how strongly people feel. In other words,

---

[8] An alternative version of this classification scheme is built on four types instead of three: direct natural, direct constructed, proxy natural, and proxy constructed (Keeney and Raiffa 1976; Parnell 2007).

[9] The stock price multiplied by the number of shares yields the company's market capitalization, but clearly I'm oversimplifying. In reality, the stock price reflects many other factors beyond the perceived value of the company.

though each member of the public could be described as supportive, opposed, or neutral, two populations may have very different intensities of support (or opposition), suggesting that the prospects of raising money for advertising on behalf of the law are also different. There is no natural attribute for this intensity of support.

Simple measurements of physical quantities are usually natural attributes when they are directly connected to the behavior of interest. For example, if I were doing a study of the weights of children of various ages, weight measurements from a doctor's scale would be a natural attribute. Of course, if I were using those same weight measurements as an indicator of some other behavior, e.g., the children's health, the same measurements could be classified as a different type of attribute. Remember, attributes are a method of classifying the relationship between data elements and the behavior of interest – they should not be applied to the elements alone.

2.4.2   Constructed Attributes

When you can observe the behavior of interest but there is no natural scale with which to measure it, constructed attributes are a good alternative. A constructed scale is specially built to capture information regarding the behavior of interest.

There are two primary methods of building constructed attributes. First, continuous constructed attributes can be built by combining other elements using a mathematical formula. Second, discrete constructed attributes can be built from tailored ordinal scales, with the scores corresponding to linguistic descriptions of different states of the behavior of interest. Building a constructed attribute is a skill in its own right – for more detail see Keeney and Gregory (2005) and Keeney (2007).

Let's continue with the stock market example. Like many traders, an analyst may be interested in the overall health of the market, but there is no obvious natural attribute available. To fill this need, traders built indices such as the Dow Jones Industrial Average and the S&P 500 index, which are both constructed attributes assembled from a computational composite of select companies' stock prices. These indices are used by traders to quickly characterize the market's overall health. This is an example of the first method of building a constructed scale.

For the intensity of public support example, consider the constructed scale in Table 2.1, where I've broken out the intensity of public support or opposition into seven levels. Note that in terms of Stevens' levels of measurement, this is an ordinal scale. In practice (and as seen in Table 2.1), the attribute levels should also have written descriptions of each level of support so unambiguous assignments could be made. If you were able to survey a population and and determine which of the seven levels best fit people's feelings, that would tell you something about the intensity of public support. This is an example of the second method of building a constructed scale.

TABLE 2.1
Constructed attribute for intensity of public support.

Label	Name	Description
A	Strong support	Vast majority of population are aware of and in favor of the proposed law. Supporters organize frequent and well-attended rallies to raise awareness. There are no rallies in opposition.
B	Moderate support	Majority of population who are aware of the proposed law are in favor of it. Most of those who are unaware, when informed of its provisions, express favorable opinions. Rallies in support and opposition are infrequent or poorly attended.
C	Weak support	Majority of population who are aware of the proposed law are in favor of it, but do not indicate a strong opinion. No rallies are organized.
O	Neutral and undecided	Majority of population have not yet formed an opinion regarding the proposed law, or are unaware of the proposal. No rallies are organized.
X	Weak opposition	Majority of population who are aware of the proposed law are opposed to it, but do not indicate a strong opinion. No rallies are organized.
Y	Moderate opposition	Majority of population who are aware of the proposed law are opposed to it. Most of those who are unaware, when informed of its provisions, express unfavorable opinions. Rallies in support and opposition are infrequent or poorly attended.
Z	Strong opposition	Vast majority of population are aware of and opposed to the proposed law. Opposition organizes frequent and well-attended rallies to raise awareness. There are no rallies in support.

Keeney and Gregory (2005) make a key point regarding the classification of attributes: "It is familiarity with an attribute and the ease of interpreting the attribute levels that distinguishes natural attributes from constructed attributes." For people who regularly observe the stock market, the composite indices are used as if they were natural attributes. In practice, there is a lot of grey area between natural and constructed attributes. The final category, however, is much easier to differentiate from the first two.

2.4.3 Proxy Attributes

Proxy attributes are used when you can't directly observe or measure the behavior in which you are interested. Instead, the proxy attribute measures a quantity that you actually have access to and that indirectly indicates the behavior of interest. Clearly this isn't an ideal situation, but sometimes an analyst has no other choice. Social scientists, who often need to use proxy attributes, call them indicators (Frankfort-Nachmias and Nachmias 1992).

The stock market indices I discussed as constructed attributes can serve as proxy attributes for concepts that don't have obvious natural attributes. For example, the overall health of the economy is a notoriously difficult quantity to measure, but it does have an indirect relationship with stock market indices. Gross Domestic Product (GDP) is another proxy attribute for overall economic health. When these proxy attributes trend positively, that is an indication that the overall health of the economy is good, but there is certainly no guarantee that economic health and the proxies will always be connected. For example, a situation could arise where the stock market and national GDP were both trending positively, but the unemployment index (another proxy for economic health) was trending negatively. In combination, these proxy attributes would not tell a simple story about overall economic health.

Another example of a proxy attribute is the Apgar score (Apgar 1953) used to assess the overall health of a newborn. Of course, there is no natural attribute for this behavior. The Apgar score combines measurements of five criteria, including pulse rate and respiration, that are correlated with but not directly representative of a newborn's overall health.

2.4.4 Commentary

Attributes are a useful classification scheme for analysts because they describe the distance between the measured quantity and the behavior of interest. The larger that distance, the greater the likelihood that some factor could intervene between the measurement and the behavior and impact the analytical conclusions. The key to using constructed and proxy attributes correctly in your analysis is to understand this limitation and think through its implications for the analytic question at hand.

Many analysts will jump at measuring whatever data is at hand without considering its connection to the behavior of interest. Instead, reflect on what sort of data you are using. Natural attributes are preferred to constructed ones, and constructed attributes are preferred to proxy ones. On the occasions that you have high precision proxy attribute elements as well as low precision natural attribute elements, it's probably best to analyze both and compare and contrast the findings. Natural attribute elements are extremely valuable!

2.5   Data Sets

Now that you understand the types of elements you will encounter, it's time to learn about the forms in which you will encounter them. Elements are usually measured and analyzed in groups that contain multiple elements. As I mentioned in Section 2.1, these groups of elements are known as <u>data sets</u>. The elements in a data set have some sort of relationship – they could be repeated measurements of the same physical quantity or measurements of a quantity at different locations, for example. One purpose of this book is to help you learn the techniques, both technical and otherwise, to analyze data sets and produce findings.

Data sets take a variety of structural forms, and their structure, combined with the characteristics of their constituent elements, influences the types of analyses that can be applied. We'll begin our discussion of data sets with lists, a type of data set with minimal structure.

2.6   Exercises

1. Classify the following elements using Stevens' levels of measurement. Explain the reasoning behind each choice.

   (a) The distance from New York to San Francisco
   (b) The number of plates in your kitchen
   (c) The name of a book on your shelf
   (d) The height of your home above sea level
   (e) The day, month, and year of your birth
   (f) Your age
   (g) Your gender
   (h) Your age, as one of these categories: child, adult, or elderly

2. Which of the elements from the previous exercise can be classified in terms of continuous or discrete data? How would you classify those elements?

3. Consider the following hypothetical situation:

   "John is an analyst for a major hotel chain and is trying to decide where to recommend constructing a new hotel. John must choose one site from five possibilities. John ranks the five sites in terms of two important factors, cost of construction and expected demand, with the best hotel site in each category being scored as a 5 and the worst as a 1. John adds the two scores and recommends the hotel with the highest total score."

   Describe what John is doing in terms of the three classification schemes in this chapter. How could John's system lead him astray?

4. In the situation described in the previous exercise, what simple methodological changes could improve John's system for choosing the best hotel site?

3

Lists

3.1 Introduction

A <u>list</u> is a common and simple data structure consisting of a group of related elements, distinguished by the fact that the order of its constituent elements can be rearranged without altering the list's information content. Note that this definition of a list excludes what are commonly known as ranked lists. Some examples of lists are the heights of the eighth graders at your local middle school, the birthdays of your friends, or the prices of all the books on your bookshelf.

The elements that make up a list should be of the same dimensionality. For example, a list could be a collection of length measurements or a collection of weight measurements, but it should not mix the two. The techniques I will present in this book all boil down to making comparisons between elements, and it doesn't make much sense to compare "2 meters" with "5 kilograms." There's no sensible way to perform any kind of comparison between the two pieces of data. This doesn't mean that a list can only be made from ratio scale elements with the same units; for example, a list of the colors of all the cars in a parking lot is a perfectly good subject for analysis, following the rules that govern categorical scales.

Some formalism will help us discuss lists in a generalized manner. A list q has n elements q_i, where i ranges from 0 to $n - 1$. Following my definition of a list, shuffling the order of the elements does not change q into a different list.[1]

---

[1]Calling the first element in the list q_0 is the zero-based array subscript convention, which I will employ throughout this book. The one-based array convention, which is equally common, labels the first element as q_1 instead of q_0. For those of you who are programmers, if the programing language you are using employs the one-based convention, be sure to adjust the indices in the equations before inserting them into your code.

3.2 Two Example Lists

In this chapter, I will analyze two different lists. Neither list has uncertainty in its measurements; I'm going to discuss uncertainty in detail in Chapter 4. All the data sets in this book can be downloaded at www.applyinganalytics.com.

3.2.1 Heights of People in a Classroom

The first list is an illustrative data set of the heights, in inches measured precisely, of 33 occupants of a middle school classroom. This will be known as the height list. There are 15 boys, 15 girls, and 3 teachers in the room (this hypothetical classroom has an excellent student to teacher ratio). In terms of the classification schemes discussed in Chapter 2, this is a continuous list that has been discretized by the finite precision of the computer variables that hold the data.[2] Clearly there is no way to practically measure the height of a person to this level of precision, but let's accept it for the sake of argument. The elements are ratio scale numbers measured with a natural attribute.

The height list, rounded to the nearest tenth of an inch, is as follows:

50.8, 48.6, 51.0, 51.8, 50.1, 46.9, 48.5, 50.7, 49.7, 50.6, 49.1,
51.2, 49.1, 49.0, 50.8, 51.7, 50.5, 51.3, 55.2, 51.5, 52.2, 52.3,
52.6, 52.9, 52.4, 51.1, 50.4, 51.6, 53.1, 51.6, 69.7, 66.2, 67.8.

This is the only example data set for which I will write out all the elements; the other data sets in this book have too many elements to print out in this manner.

The first step in analyzing this data set is to sort it,[3] from smallest to largest element:

46.9, 48.5, 48.6, 49.0, 49.1, 49.1, 49.7, 50.1, 50.4, 50.5, 50.5,
50.7, 50.8, 50.8, 51.0, 51.1, 51.2, 51.3, 51.5, 51.6, 51.6, 51.7,
51.8, 52.2, 52.3, 52.4, 52.6, 52.9, 53.1, 55.2, 66.2, 67.8, 69.7.

My list element notation will refer to this sorted ordering of the list, e.g., $q_{10} = 50.5$ in. It's much easier to identify trends in a sorted list than in an unsorted one. The range of the data set, defined as the smallest and largest elements, is now easily recognizable. Elements with values far from the other members of the data set, known as outliers, also stand out better when a list is sorted.

Some conclusions jump out from an initial examination of the elements in

---

[2]The height list was constructed from three different normal distributions (a topic I will discuss later in this chapter): the boys' distribution has a mean of 50 in., the girls' 52 in., and the teachers' 69 in. Don't use this data set for anything other than testing code, since the data are made up.

[3]Sorting lists is a matter of some art, and which method is most efficient actually depends on the original order of the list. For more information, see Chapter 8 of Press et al. (2007), Knuth (1998), and Deitel and Deitel (2001).

the sorted height list. Most of the elements cluster around 51 in., with three large outliers around 67 in. There is a gap from 56 to 66 in. where there are no heights measured. The range of the data set is from 46.9 to 69.7 in. The ratio of the absolute value of the largest (q_{32}) to the smallest (q_0) elements is about 1.5; this quantity is known as the dynamic range (see inset).

Dynamic Range

The dynamic range of a data set is calculated by taking the ratio of the element with the largest absolute value to the element with the smallest absolute value. It should be calculated neglecting 0 as a data value, because that would cause the dynamic range to be infinite – use the next smallest value instead. The dynamic range of a data set measures the extent to which the elements vary, but it does so in a way that is independent of the magnitude of the element values. In other words, the fact that the largest element is 1.5 times the smallest element can often be a more useful piece of information than the fact that the largest element is 22.8 in. bigger than the smallest element (as in the height list). As I will discuss throughout this book, the dynamic range strongly influences the analytic and visualization techniques that work best for a given data set.

3.2.2 Ages of U.S. Citizens

The second data set is a list of the ages of all the inhabitants of the United States in November 2007. This will be known as the age list. This list was created by the Population Estimates Program in the Population Division of the U.S. Census Bureau. The Census Bureau publishes a monthly estimate of the population by age; if the estimate stated that there were 4 million people between the ages of 1 and 2, I inserted 4 million 1's into the list. I then randomized the order of the elements to better recreate what the data would look like if it were gathered through a census.

In terms of the classification schemes from Chapter 2, this is a discrete list with 303,259,991 elements. The elements are ratio scale numbers measured with a natural attribute. As an artifact of the way the list was created and how the data were disclosed by the survey's authors, all ages greater than 100 yrs appear in the list as elements with a value of 100 yrs. This inaccuracy affects a small number of the entries in the list, but is important to keep in mind regardless. In general, real data sets often have similar limitations; perfectly gathered data is hard to come by.

From an initial examination of the list it is apparent that the elements range from 0 to 100 yrs, and have a dynamic range of a factor of 100. Ages toward the high end of the range are relatively rare compared to the low end and midrange values.

3.3   Elementary Analyses

Every student is taught to calculate the mean, median, and mode in elementary school – that's why I call them elementary analyses. The procedures for calculating these statistics, processes that are more formally known as algorithms, are simple, but there is skill in interpreting their outputs. The presence of outliers and the structure of the data should influence an analyst's choice of technique and how the outputs are communicated. Which measures best suit each circumstance?

Often one of the first things an analyst needs to know about a data set is the value of a "typical" element because that will aid in speaking broadly about the data set's content. The typical value doesn't have a precise definition; if it did, I could write an algorithm for it! In fact, choosing a technique to estimate the typical value is a subjective decision, and I will discuss why some choices are better than others. In general, you want to use a statistic that isn't overly reliant on the value of any one element because these measures would be susceptible to errors in that single measurement; statistics that aren't unduly influenced by a small group of elements are called robust. In the discussion that follows, I will demonstrate some of the pitfalls involved in finding a typical value for a data set.

To aid in discussing the elementary analyses, I will introduce an alternate view of a list's data content, known as the frequency distribution. The frequency distribution, $N(q)$, is a function that returns the number of elements that have a particular value, q. For lists, the frequency distribution contains all of the information contained in the list form of the data, though for more complex data structures information about the elements' locations is lost in the transition. The frequency distribution is more straightforwardly constructed for data structures that consist of elements measured with discrete scales, but it can also be built for data sets consisting of elements measured with continuous scales. Though the distribution function can take as an input quantities that have dimensions (such as meters or years), the output N has no units. You can think of the distribution function as supplying the answer to a question like: "How many elements in the data set have a value of 3 meters? 5." The frequency distribution of a data set is sometimes called simply the data set's "distribution" as shorthand. For a discussion of how to calculate the frequency distribution, see the inset, and for a discussion of how to visualize it, see Section 3.4.

The three elementary analyses reduce all the information in a large data set to a single number. This is, simultaneously, the power and the problem inherent in the elementary analyses. You're trying to describe a data set, which has many layers of detail in its frequency distribution, with a single piece of information. For better and for worse, this hides most of the data set's features.

Frequency Distribution

To calculate the frequency distribution for a data structure that contains elements measured with a discrete scale, first find the lowest and highest element values, which I will call q_L and q_H. For each allowed value of q between these endpoints:

$$N(q_L \le q_j \le q_H) = \sum_{i=0}^{n-1} (q_i \text{ eq } q_j), \tag{3.1}$$

where the "eq" operator returns 1 if the two elements are equal and 0 if they are not. This algorithm cycles through q and totals up the number of elements with value q_j. Technically, this new construct N is a 1-dimensional data structure (see Chapter 5), not a list.

A note for programmers: algorithmically, building N from the the definition in Equation 3.1 can sometimes be a bit slow when working with a large data set. An alternate algorithm is to cycle through the list, adding an entry to the frequency distribution whenever you encounter a value that hasn't appeared previously. If the value has appeared before, increment the counter associated with that value. When you come to the end of the list, sort the values you have encountered.

To calculate the frequency distribution for a data structure containing elements measured with a continuous scale, the elements must be placed into bins. The analyst has freedom to choose a binning structure that is convenient for the particular analytic questions of interest. Each element must be placed into one and only one bin.

3.3.1 Mean

Calculating the <u>mean</u>, μ, (also known as the average or the first moment) of a list q is straightforward. It is simply:

$$\mu = \frac{\sum_{i=0}^{n-1} q_i}{n}. \tag{3.2}$$

For readers who aren't familiar with mathematical notation, the right-hand side of Equation 3.2 is the sum of the values of all the list's elements, starting with q_0 up to and including q_{n-1}, divided by the number of elements.

The mean is not necessarily equal to the value of any of the elements in the list; if the gaps between element values are large, the mean can fall into one of these gaps. This fact is occasionally important when using the mean as a proxy for a typical element in the list, such as in the case of a list made of elements measured with a discrete scale. For example, when the mean of a list of 1's and 0's falls somewhere in between the two values, the mean may not be telling you what you need to know.

Although the algorithm for the mean is uncomplicated, its interpretation can be quite difficult because the mean hides many of the data set's features. Outliers can <u>dominate</u> the mean, meaning they can exert disproportionately strong influence on the algorithm's output. Let's illustrate the influence of outliers on the mean and the concept of dominance with a notional example. Imagine a sample of 50 people, 49 of which have an income of $50,000 each and 1 who has an income of $100 million; the average income of the sample is $2 million. Is this really a meaningful typical value? The income of the 49 people could have been $0 and it would have barely changed the mean, because the single outlier dominates the calculation. The actual mean of American incomes is influenced by a weaker version of this effect.

A simple way to estimate the influence of outliers on μ is to calculate the mean of q excluding the top and bottom 5% of elements. How much does the mean change with this modified data set? If there is a large shift in the mean, you will likely need to find a different technique to determine and communicate the typical value of q. If the mean is approximately unchanged, the calculation is likely dominated by the central values of q, which means that the statistic is likely robust.[4]

Understanding the influence of outliers does not constitute blanket permission to discard them from your data set. Consider, for example, a financial analyst modeling the daily fluctuations of the stock market indices. She has a list of daily market index changes by percentage, which cluster closely together except one element corresponding to October 19, 1987, which contains a 22% drop. Is the analyst justified in throwing out this point and making decisions based on the new μ? It depends. If the data on that day were collected using a different method than the one employed on other days, then it could be justified. After all, comparing apples and oranges is not helpful. However, it turns out that this element corresponds to the stock market crash known as Black Monday. Throwing out this element would result in the loss of a very important piece of information! In Section 3.8, I will discuss a technique that will help you make the decision of whether an element can justifiably be thrown out.

For lists made up of elements measured with discrete scales, it is often easier to use the frequency distribution to perform calculations rather than the list form of the data. To calculate the mean from the frequency distribution, use:

$$\mu = \frac{\sum_{q_j=q_L}^{q_H} q_j N(q_j)}{n}, \tag{3.3}$$

where the sum runs over all the possible values of q between q_L and q_H. Using the frequency distribution is particularly helpful when the original list is large enough to constrain computational resources and when the values in q repeat. A good rule of thumb is to use the frequency distribution when the number

---

[4]Note that this technique can't be used to estimate the influence of outliers for all data sets. For example, if a list consists of two clusters of elements, outliers may be found between the clusters and will not be part of the top or bottom 5% of elements. Therefore, examine the frequency distribution and use your judgment before applying this technique.

of possible values of q is considerably smaller than the number of elements n, i.e., when there are many repeated values in q. The age list is an example of a data set with these characteristics. For data sets measured with continuous scales, the frequency distribution should not be used to calculate the mean because the binning process has altered many of the elements' values.

The mean of the height list is 52.5 in. In my earlier inspection of the data set it was apparent that most of the points clustered around 51 in., and μ isn't far from that value. However, if I remove the three tallest and shortest heights and recalculate, the mean falls to 51.3 in. The drop is clearly due to removing the teachers from the sample; they function as outliers in this data set, though their influence is not large enough to dominate the mean. If your goal is to calculate the typical height of a *child* in the class, you could justify removing these outliers. On the other hand, if you are trying to design a system that depends on the typical height of a *person in the classroom*, the mean is a misleading statistic to use. It would be much better to design the system with two separate populations in mind: one with a height of 51 in., and another at 69 in.

The mean of the age list is 36.9 yrs. Outliers in the age list are unlikely to have a strong influence on the mean for several reasons. First, in this list outliers are only possible as large positive numbers (no one is younger than 0 yrs old). Second, even the healthiest people rarely live much longer than 100 yrs. Third, due to the manner in which the data were gathered and recorded, even people older than 100 yrs appear in the list with their ages truncated to 100 yrs, limiting their impact.

3.3.2 Median

If the list q is sorted so its elements are in order, either from smallest to largest or largest to smallest, the median γ is the middle value in the list; a point where there are an equal number of elements above and below it. Note that there is no agreed upon symbol for the median; I've chosen γ for convenience. Like the mean, the median can also be used to determine a data set's typical value.

The algorithm for calculating the median distinguishes between data sets with an odd number of elements, where the median follows γ_{odd}, and those with an even number of elements, where the median follows γ_{even}. In sorted lists where n is an odd integer, the median is simply:

$$\gamma_{\text{odd}} = q_{(n-1)/2}. \tag{3.4}$$

For sorted lists where n is an even integer, take the mean of the middle two elements:

$$\gamma_{\text{even}} = \frac{q_{(n/2)-1} + q_{n/2}}{2}. \tag{3.5}$$

For large data sets, it is unusual for the results of your analysis to be affected if you instead take either $q_{(n/2)-1}$ or $q_{n/2}$ to be γ_{even}. In other words, if the median is sensitive to replacing the average of the middle two elements with one of the elements alone, that's an indication that the median is not robust and you may want to look to some other technique for determining a typical value.

The median is generally far better than the mean at characterizing a typical value of the data set because it is only weakly influenced by outliers. The median of our notional list of 50 incomes (discussed in Section 3.3.1) is $50,000, which matches my intuition of what a typical value should be; the outlier had no impact on this statistic. The median is particularly well-suited for application to data sets measured with ordinal scales, where in many cases it is mathematically inappropriate to calculate a mean. For discrete data sets, especially those with irregular and/or large gaps between data values, some caution is required in using the median since small changes in the data set can produce wildly different results – in these cases, the median can behave in a way that is not robust.

When analyzing a data set, you can gain insight by comparing the calculated mean and median. How different are the two statistics? If they are significantly different, why are they different? If they are similar in value, why are they similar?

Occasionally, analysts will use finer divisions, constructed similarly to the median, to describe how the elements in a list are distributed. Two common examples are quartiles and deciles. In a sorted list, the first quartile ranges from the smallest element in the list to the element q_i where $i = n/4 - 1$. The second quartile ranges from the next element in the list, $q_{n/4}$, to the median. The third and fourth quartiles follow in the same manner. Deciles are structured similarly, with the list split into 10 equally populated segments, rather than 4. The finer the divisions, the more information about the frequency distribution these techniques provide.

For the sorted height list, which has an odd number of elements, $\gamma = q_{16} = 51.2$ in. Note how the median approximately matches the value of the mean after the outliers have been removed (51.3 in.), again demonstrating the influence of the outliers on the mean. For the height list, the median is a better representation of a typical height in the data set than the mean.

For the sorted age list, which also has an odd number of elements, $\gamma = q_{151629995} = 36$ yrs. This is fairly close to the mean (36.9 yrs), with μ being slightly larger due to the shape of the frequency distribution (i.e., there are no people younger than zero years old). In the age list, there isn't much difference between the mean and the median, and which one you use in your analysis shouldn't have a big impact.

3.3.3   Mode

The mode is defined as the value of a data set that appears most frequently. For a data set measured with a discrete scale, finding the mode is a two-step process. First, solve for the frequency distribution, then find the value q_j where

$N(q_j)$ is at its peak. This value is the mode. The mode can be multivalued if N has multiple peaks of the same height. When that happens, don't average the values together; just report all the peaks.

For a data set measured with a continuous scale, or a data set measured with a discrete scale in which there are many more possible values than elements, an analogous technique to the mode exists. First, bin the data, then find the bin with the largest population. In this case, the "mode-like" quantity will depend on the placement and width of the bins.

For a data set consisting of elements measured with a categorical scale, the mode is one of the few mathematical techniques at your disposal. It is simply the category with the largest number of elements.

Frankly, the mode is the black sheep of the elementary analytic techniques family, but there are occasions when it is a useful statistic. For example, examining the difference between the mode and the median of the data set tells you about the shape of the frequency distribution. Granted, with modern computers it's not difficult to simply plot the frequency distribution itself, a technique discussed in the next section.

The height list is practically continuous, so I'll determine the "mode-like" quantity by following a binning procedure – this will essentially convert the list into a data set measured with a discrete scale. One way to bin the elements is to round all the entries down to the nearest integer. This operation is called the "floor" of a number. For example, the floor of 46.2 is 46.[5] After the data are binned in this way, the peak in the frequency distribution appears at 51 in., where $N(51 \text{ in.}) = 9$. Of course, choosing a different bin pattern may result in a change in the "mode-like" quantity. This quantity is near the median (51.2 in.) and slightly lower than the mean (52.5 in.), consistent with my finding that the height list has most of its points clustered near 51 in. with a few outliers at large height.

There is no need to bin the age list since the data are already measured with a discrete scale and each possible value occurs many times. The mode of this data set is 47 yrs, with $N(47 \text{ yrs}) \approx 4.7 \times 10^6$ (the $\approx$ symbol means approximately equal to). This is larger than the median (36 yrs) and the mean (36.9 yrs), suggesting a frequency distribution that has lower counts at higher ages.

3.4 Additional Techniques

All but the simplest data sets will not be fully characterized by the mean, median, and mode; most data sets are far too complicated to be described by

---

[5]Contrast this with the "ceiling," which is the number rounded up to the nearest integer. The ceiling of 46.2 is 47.

just three numbers. Adding a few more techniques to our toolbox will improve our ability to analyze lists and other data sets.

3.4.1   Histogram

Plotting a histogram is an effective way to examine a data set's frequency distribution. In its simplest form, a histogram is a plot of the count of the number of list elements q_i within a series of data value ranges. Generally, histograms are plotted with data values on the horizontal axis and counts on the vertical axis. Peaks in the histogram correspond to common data values, and valleys to data values that appear infrequently in q. Each data element q_i should fall into exactly one bin, no more and no fewer. You'll want to make sure not to double count elements that fall exactly on the divider between two bins; a good sanity check is to total up all the bars in the histogram and make sure the number matches the number of elements in the list, n.

There is some subjectivity to choosing the size and setup of the bins for a histogram. Ten bins is generally a good number to shoot for, though I've certainly seen cases where more finely binned histograms were useful. You'll usually want all your bins to be the same width. Sometimes you may want to make an exception to consistent width for the highest and lowest bins, because you may need to take extra care to catch all the outliers with high and low values. A histogram that simply excludes outliers by not plotting them can be misleading. One option is to define your highest bin as the number of elements larger than some value, but if you do this you'll want to label that bin differently to avoid confusion.

For data sets measured with continuous scales, you will need to choose the upper and lower data values of each bin. For data sets measured with discrete scales, if there aren't many possible values, each can get its own bin, but if there are many values then they may need to be grouped. For example, the bin width for the discrete age list, which has 101 possible values, could be ten years, e.g., 0–9 yrs for the first bin, 10–19 yrs for the second, etc., with the final bin being 100+ yrs.

Sometimes other considerations will influence your choice of binning scheme; for example, if you find yourself analyzing a list with uncertainty in its data values, the typical uncertainty can be a natural scale for bin width. If your data set contains both positive and negative values, you should do something intelligent with the zero value, such as placing it centered in a bin.

When you've finished plotting the histogram, take a close look at the shape of the result. Is the frequency distribution lopsided? How centrally concentrated are the elements? How many peaks are there in the frequency distribution? Can you explain the origin of each peak? If it is relatively flat, why is it flat? Are there valleys in the frequency distribution? What are their origins?

Figure 3.1 is a histogram for the height list. It provides graphical evidence for the frequency distribution I described when I calculated the mean, median, and mode. The bins are structured so that the bar corresponding to 46

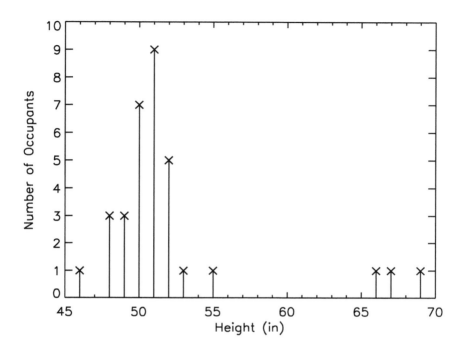

FIGURE 3.1
A histogram of the heights of the 33 inhabitants of an elementary school classroom. The point at 46 in. corresponds to the number of occupants with height h that satisfies $46 \leq h < 47$ in., and so on for each of the other bins.

in. contains all the elements between 46.0 and 46.999... in., and so on for each bin, identical to the binning described in Section 3.3.3. This histogram shows a roughly bell-curve (normal) distribution (see Section 3.5) centered around 51 in., and three outliers between 65 and 70 in.

The age list histogram is plotted in Figure 3.2. Here the data have been binned in ten-year intervals. The population in each bin stays roughly constant with increasing age until reaching a break point around the 60–69 yr bin. At that point, the population falls off fairly steeply until the number of Americans older than 100 yrs is about a factor of 1000 smaller than the population in the younger bins.

A few tips on plotting histograms are worth mentioning. It must be clear which bar corresponds to each bin; this can be particularly difficult when your histogram has many bins. In Figure 3.1, the vertical lines connect the x-shaped points to the x-axis and give the frequency distribution a defined shape that is easy for the eyes to interpret. Notice that in both of these histograms, the data points are not connected with horizontal lines. As a rule of thumb, when

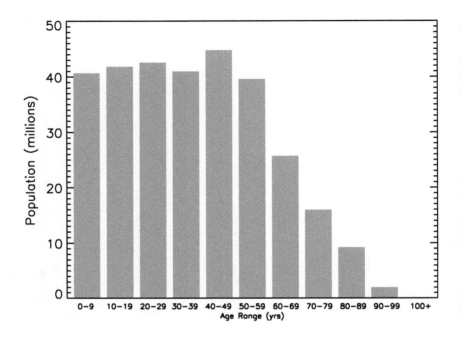

FIGURE 3.2
A histogram of the November 2007 American population binned in ten-year increments.

the data are binned over the x-axis, do not draw a connection unless the complexity of the figure demands it. For example, in the case of Figure 3.1 there is no bin corresponding to the value 50.5 in., so why should you draw a line between the points at 50 and 51 in?

Analysts have developed many different variations on the simple histograms discussed in this section. These variants are functionally similar to simple histograms, though they often convey select pieces of additional information. Just a few examples of the many variants include:

- Histograms rotated 90°, so that the bars extend horizontally instead of vertically.

- Histograms in which the bars consist of stacked shaded or colored smaller bars, where the shades and colors correspond to different element categories. With some additional information, the age list could be plotted in this way, with stacked bars of different shades corresponding to the number of each group that is male or female.

• Histograms where the horizontal spacing between the bars varies, so some bars are grouped together for ease of comparison.

3.4.2   Stem and Leaf Diagrams

Stem and leaf diagrams are similar in structure to histograms, but they convey more information. This is both a blessing and a curse, as sometimes the extra information aids analysis, and at other times it distracts from the behavior you are trying to illustrate. Personally, I like to use stem and leaf diagrams as tools for my own analysis, but I use histograms when presenting findings to people who aren't as familiar with the data as I am.

The easiest way to learn how to construct a stem and leaf diagram is from an example. Table 3.1 depicts the stem and leaf diagram for the height list. The stems are the numbers to the left of the solid vertical line and the leaves are the numbers to the right. In this example, each leaf represents one element in the height list. For example, the three smallest points in the height list are 46.9, 48.5, and 48.6 in. The smallest of these elements appears as a leaf off of the "46" stem, and the next two as leaves off of the "48" stem. Notice that the "49" stem has no leaves. The total number of leaves in this figure is equal to the number of elements in the list, n.

Compare the vertical shape of the stem and leaf diagram to the horizontal shape of the histogram in Figure 3.1. Notice how the stem and leaf diagram conveys the features of the frequency distribution, but does so with added information. In what contexts would you use each of the two diagrams?

When constructing a stem and leaf diagram of your own, you shouldn't necessarily use all the decimal places of precision available. It is perfectly acceptable (in fact, it's recommended) to round the elements, such that the leaf consists of only one digit. This digit should not necessarily correspond to the first one to the right of the decimal point. See, for example, the stem and leaf diagram for the age list (Table 3.2), where the stems increase in 10 year increments. Similarly, each leaf doesn't need to correspond to a single element. In the age list stem and leaf diagram, each stem represents 10^6 people of that age. If there were fewer than 10^6 people of that age, I didn't add a leaf to the diagram. This accounts for the differences in shape between the stem and leaf diagram and the histogram (Figure 3.2) at high ages.

Choosing the stem structure for a stem and leaf diagram is a similar process to choosing the binning structure for a histogram. You don't want all of your leaves to come from one or two stems, but neither do you want to have nearly as many stems as you do leaves. Fortunately, you have a lot of freedom in setting up the stem structure. Emerson and Hoaglin (1983) describe how stems can be split, even if they have the same first digit, as shown here using the "10" stem for the age list:

TABLE 3.1
Stem and leaf diagram for the height list, in inches.

```
46. | 9
47. |
48. | 56
49. | 0117
50. | 1455788
51. | 012356678
52. | 23469
53. | 1
54. |
55. | 2
56. |
57. |
58. |
59. |
60. |
61. |
62. |
63. |
64. |
65. |
66. | 2
67. | 8
68. |
69. | 7
```

```
1* | 000111222233334444
1○ | 55556666777788889999
```

Here, "1*" is the stem for elements with values 10 through 14, and "1○" is the stem for elements with values 15 through 19. You can imagine using additional symbols if you'd like to split the stems further. Keep in mind that as with histogram bins, it's best to build the stem structure so that each stem has equal range.

One last tip: when constructing your own stem and leaf diagrams, make sure you use a fixed width font. If the numbers in your font are of variable width, the size of the rows of leaves will not be proportional to the number of entries.

Stem and leaf diagrams are easy to construct by hand for small data sets, so they were commonly used before computers were widely available for data analysis (Mosteller and Tukey 1977; Tukey 1977; Emerson and Hoaglin 1983). Back in the late '70s and early '80s, analysts developed an intricate shorthand to mark important features of the data on these diagrams that is fascinating to look at in retrospect.

TABLE 3.2
Stem and leaf diagram for the age list, in years. Each leaf represents 10^6 people of that age.

0	00001111222233334444555566667777888999
1	000111222233334444555566666777788889999
2	0000111122223333444455556666777788889999
3	000011112223334445555666677778888999
4	0000111122223333444455556666777788889999
5	000011112222333344455556666777788899
6	00011223344556677889
7	0123456789
8	0123
9	
10	

3.4.3   Variance

Variance is a measure of a data set's scatter around its mean.[6] Variance is just one of many statistics designed to describe this aspect of a data set's frequency distribution, known as its dispersion. Essentially, the variance helps an analyst determine where a particular data set falls between two extreme cases: all the elements located very close to the mean or spread out with no central clustering whatsoever.

There are two common formulae for calculating the variance s of a data set. The first is given by:

$$s_n = \frac{\sum_{i=0}^{n-1} (q_i - \mu)^2}{n}, \tag{3.6}$$

and the second by:

$$s_{n-1} = \frac{\sum_{i=0}^{n-1} (q_i - \mu)^2}{n-1}. \tag{3.7}$$

Both formulae sum the squared differences between each element and the mean. The two formulae differ only in the fractions' denominators, and there they differ by a factor that makes only a small difference when n is large.[7] The units of the variance are the units of the list elements, squared. Since s is the sum of a squared quantity, it is always greater than or equal to zero. Again, if your results depend strongly on which formula you choose, that's generally a sign either that you've made a mistake or that the variance is not a robust statistic for your data set.

Variance, by itself, is a difficult statistic to develop an intuition for because its units don't match the units of the list elements. The standard deviation,

---

[6]Technically, variance is also known as the second central moment.
[7]The two formulae differ for technical reasons related to bias.

discussed in the next section, is an improvement on the variance for this reason. For what it's worth, let's go through a simple interpretation of the variance. From Equations 3.6 and 3.7, you can see that the further an element is from the mean, the more it adds to the variance. This happens regardless of whether the element is larger or smaller than μ, because the difference is squared so any difference will add to s. For this reason, data sets that have larger s tend to be more spread out from the mean. Dividing by n (or $n-1$) normalizes s so that data sets that have more elements don't necessarily have larger variances.

The variance can be a misleading statistic if you use it blindly. Imagine a list with a bimodal distribution; the elements cluster around two different values. The mean for this list will fall somewhere between the two peaks in the frequency distribution, therefore elements located near either peak will contribute a relatively large value to the variance. This will occur even if the elements are clustered very tightly around those two peaks. The variance will be large due entirely to the bimodal nature of the frequency distribution, and nearly independent of any intrinsic scatter of the elements around those peaks.

Another common situation in which the variance can lead you astray is in the case of large outliers. Because elements contribute to the variance by their squared difference from the mean, the variance can be dominated by a small number of the elements in the list if they fall far enough from the mean. Again, parallel to the recommendation in our discussion of the mean, it is a good idea to recalculate the variance for comparison's sake after removing the top and bottom 5% of elements from the list.[8] Consider our notional example of the incomes of 50 people from Section 3.3.1. The variance of this list is entirely due to the presence of the one outlier ($s_n = 2 \times 10^{14}$ dollars^2).

For the height list, $s_n = 26.3$ in.^{2} and $s_{n-1} = 27.2$ in.^{2}. Contrast this with the variance calculated after removing the top and bottom 3 elements (this is more than 10% of the total number of points, but bear with me): $s_n = 1.8$ in.^{2}. Clearly, the three outliers in this data set have an enormous impact on the variance.[9]

For the age list, $s_n = 511$ yr^2. I've already discussed how the construction of this list limits the potential influence of outliers by setting the maximum element value to 100 yrs. I can check the impact of this truncation by changing every element with a value of 100 yrs. to 110 yrs. and recalculating the variance (this is a "reasonable worst-case" scenario). This shifts the variance upward by less than half a year-squared, a relatively small effect.

Personally, I find it difficult to develop an intuition for the variance so I

---

[8]For readers who are coding these examples, remember that you'll want to recalculate μ when you do this as well.

[9]Readers who are paying close attention will remember that the height list was actually drawn from a trimodal distribution. By cutting out the three largest heights, I have eliminated all of the contributions from one of the distributions. The two remaining distributions with a smaller mean correspond to the heights of the boys and girls. These distributions are strongly overlapping, so the adjusted variance is not inflated.

hardly ever use it. Because the units of the variance are not the same as the data, there is no natural scale for comparison. Clearly, taking the square root of the variance is the next step.

3.4.4 Standard Deviation

The standard deviation, σ, is another technique to describe the dispersion of a data set. It is straightforwardly related to the variance and, like the variance, it has two common versions:

$$\sigma_n = \sqrt{s_n} = \sqrt{\frac{\sum_{i=0}^{n-1} (q_i - \mu)^2}{n}} \tag{3.8}$$

$$\sigma_{n-1} = \sqrt{s_{n-1}} = \sqrt{\frac{\sum_{i=0}^{n-1} (q_i - \mu)^2}{n-1}}. \tag{3.9}$$

In short, the standard deviation is simply the square root of the variance. Be aware of the difference between the standard deviation and the variance, as unfortunately many people use these terms interchangeably. The units can be a good clue as to which statistic is being used, but some analysts aren't terribly careful about putting the right units on their numbers.

The basic interpretation of σ is similar to that of s. Data sets with higher σ are less clustered around the mean. However, the standard deviation tends to be a more useful statistic than the variance because its units are the same as those of the list elements. This means that you can make informative comparisons between σ and quantities related to the distances between elements, such as how far some element is from the mean.

When you are analyzing a data set whose frequency distribution is bell-curve shaped, σ is straightforwardly connected to the width of the distribution (see Section 3.5). For several other common distributions, there are well known relations between σ and the parameters of the distribution (see Bevington and Robinson (2003)). If the underlying distribution is not one of these standard forms, all bets are off as to how σ relates to the shape of that distribution, but it is generally still a useful statistic to examine.

The name "standard deviation" causes a variety of problems; unfortunately, we're stuck with it. In general, it is only when the frequency distribution is bell-curve shaped that σ can be interpreted as a typical distance from the mean. Similarly, there is a widely known rule of thumb that 68% of elements fall within 1-σ of the mean (and so on for larger multiples of σ) – this rule only applies to the normal distribution, and is misleading for other distributions. Calculating and presenting σ for an arbitrary frequency distribution like the age list is a delicate analytic communication challenge, because many people believe they know how to interpret σ but in fact they are likely to misuse it. If you present this statistic, be sure to offer some guidance as to what intuition it gives you.

Since the standard deviation is just the square root of the variance, it is not surprising that σ and s have some of the same weaknesses. The standard deviation is just as meaningless when applied to a bimodal distribution and outliers can have a large influence.

For the height list, $\sigma_n = 5.1$ in. If I remove the top and bottom three elements (the three outliers), $\sigma_n = 1.3$ in., demonstrating again the influence of outliers on techniques that rely on squared differences. Looking at the histogram in Figure 3.1, σ_n is about half the width of the thick part of the distribution at 51 in. I'll discuss the relationship between σ_n and the width of a bell-curve distribution further in Section 3.5.

For the age list, $\sigma_n = 22.6$ yrs. Remember, I found the mean of this distribution to be 36.9 yrs. Examining the population distribution in Figure 3.2, it isn't clear what feature σ corresponds to since the shape of the curve doesn't correspond to a standard form. That happens sometimes; a statistic will not be useful for every data set you apply it to. For this reason, I likely would not present σ for the age list as part of my analysis.

A term of art that has made it into the culture-at-large is describing an element as an $X\sigma$ measurement, where X is some number. This is just a fancy way of describing the distance that a particular element is from the mean in multiples of the standard deviation. For example, for the height list ($\mu = 52.5$ in., $\sigma_n = 5.1$ in.) the person who is 67 in. tall corresponds to a 3σ measurement. People will use this language regardless of the shape of the data's frequency distribution, even though it only really makes sense for a bell-curve shaped distribution (which, of course, the height list is not). Unfortunately, I'm afraid we're stuck with this term of art.

Finally, another common use of σ is in describing the significance of a result. This is different from describing an element's distance from the mean; I'll discuss this idea further in Section 3.7.

3.4.5   Mean Absolute Deviation

The variance and the standard deviation depend on the sum of the squares of the differences between each element and the mean; for some data sets, this sum can be dominated by a few outliers. If you need a statistic for the scatter around the mean that is less strongly influenced by outliers, consider the following:

$$\tau = \frac{\sum_{i=0}^{n-1} |q_i - \mu|}{n}. \tag{3.10}$$

The quantity τ is known as the <u>mean absolute deviation</u> (or, alternatively, the average absolute deviation). There is no agreed upon symbol for this statistic, so I've chosen τ for convenience. Like the variance and standard deviation, τ is a sum of positive quantities so it is always greater than or equal to zero. Like σ, τ has the same units as the elements of q, and it can be compared to features in the distribution. Also similar to s and σ, the basic interpretation of τ is straightforward. Data sets with larger values of τ are less clustered around the mean.

For the height list, $\tau = 3.0$ in. If I remove the top and bottom three points as before, $\hat{\tau} = 1.0$ in. The outliers still influence the absolute residual sum, but at least the statistic doesn't shift by an order of magnitude, like it did with the variance. The impact of the outliers is also reduced compared to the standard deviation.

For the age list, $\tau = 19.0$ yrs. As you would expect for a list with few outliers, $\tau \approx \sigma_n$. Still, the irregular shape of the frequency distribution prevents us from attributing a straightforward meaning to either the mean absolute deviation or the standard deviation.

3.5   The Normal Distribution and the Height List

In Figure 3.1, I plotted a histogram of the height list's frequency distribution and in a footnote in Section 3.2.1, I mentioned that the elements in the height list were derived from three normal distributions. What exactly is a normal distribution and how did I use it to build the height list?

First, I need to define what I mean by a distribution. A distribution is a function that tells you how often a measurement of some quantity yields one value versus another. I've already defined the frequency distribution, which is a subtype of distribution that returns the count of each measurement value. A probability density function (see inset), on the other hand, is a type of distribution that returns the likelihood of a particular value being measured. For a given data set, the frequency distribution and the probability density function differ by a factor of the number of elements.

Probability Density Function (pdf)

A probability density function, $p(x)$, is encoded with the relative likelihood that a draw from that distribution will return a particular value x. Therefore, when an element is randomly drawn from a pdf, you're more likely to draw values from where $p(x)$ is relatively large than from where it is relatively small. A pdf can be used to calculate the probability P that an element, when drawn randomly from that distribution, falls in between x_1 and x_2:

$$P(x_1, x_2) = \int_{x_1}^{x_2} p(x)\,dx. \tag{3.11}$$

Pdf's should integrate to 1, as shown in Equation 3.13, because the chance that a value drawn from a pdf falls between $-\infty$ and ∞ should be 100%.

Sometimes, distributions have familiar shapes – that's where named distributions come in. Named distributions can appear in frequency, probability density, or some other distribution form. Though in this book I will only discuss the normal distribution, there are many named distributions specified in the literature. Some of the most common include the Poisson, binomial, beta, and Boltzmann distributions.

The normal distribution (also known as the Gaussian or bell-curve) is a named distribution widely used throughout analytics. The 1-dimensional version is given by:

$$p_{\text{normal}}(x) = \frac{1}{\sigma\sqrt{2\pi}} e^{-\frac{(x-\mu)^2}{2\sigma^2}}, \tag{3.12}$$

where σ is the standard deviation of the Gaussian centered on μ. Because the distribution is symmetric around μ, it is easy to see that μ is the mean of the distribution. It's also easy to see that the function peaks when $x = \mu$, since for every other value of x, the exponential has a value smaller than 1.

It's worth plotting a few different versions of the normal distribution to see how changing σ affects the shape of the curve (Figure 3.3). The figure demonstrates that the Gaussian widens and its peak shrinks as σ grows. Despite their different shapes, the area under each of these curves is 1. It turns out that there is a simple relationship between the width of a Gaussian and its standard deviation. The width of the distribution, measured where the curve crosses half the height of the distribution's peak, is approximately 2.35σ. This is commonly known as the full-width half-maximum, or FWHM. Changing μ (not shown in the figure) would leave the height and width of the curve the same, but shift the curve leftward or rightward on the x-axis.

The particular choice of normalization used in Equation 3.12, while not the only one you will run into, has the nice property that it integrates[10] to 1:

$$\int_{-\infty}^{\infty} p_{\text{normal}}(x)dx = 1. \tag{3.13}$$

The property described in Equation 3.13 also means that $p_{\text{normal}}(x)$, when used with this particular normalization, is a pdf.

Drawing an element from a pdf can be accomplished using a random number generator coupled with some mathematical manipulations (see the transformation method, also known as inverse transform sampling, in Press et al. (2007)). Before computers were widespread, many textbooks had lists of random numbers in their appendices, but nowadays Gaussian and uniform random number generators are built into most modern computer languages. In the case of a Gaussian pdf, a random draw is most likely to result in an element near the mean, since that is where the distribution peaks and most of the area under the curve is located. However, the larger σ gets and thus the

---

[10]Let me briefly explain what the "$\int$" symbol means to readers who are unfamiliar with it. The "$\int$" signifies integration, meaning it returns the total area under the function that follows the symbol. The dx specifies that the sum should occur over the x dimension, and the numbers to the lower right and the upper right of the integration symbol give the limits for that sum. So, the left-hand side of Equation 3.13 tells me to calculate the area under $p_{\text{normal}}(x)$ from $x = -\infty$ to $x = \infty$.

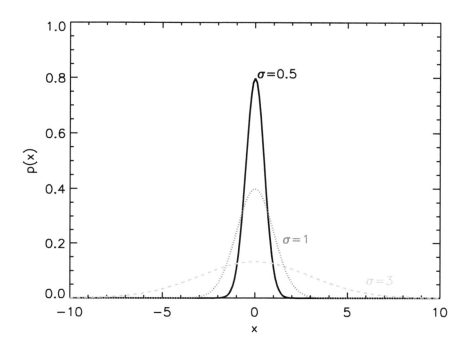

FIGURE 3.3
Three normal distributions, all with $\mu = 0$. The solid black line has $\sigma = 0.5$, the dark grey dotted line has $\sigma = 1$, and the light grey dashed line has $\sigma = 3$. Notice how increasing σ causes the peak to fall and the width to increase.

flatter p_{normal} gets (as shown in Figure 3.3), the more likely values far from the mean (as measured in x) are to be drawn.

So, how did I generate the height list using the normal distribution? All I did was create three normally distributed pdfs with different values of μ and σ: one for the boys, one for the girls, and one for the adults. I drew 15 random elements each from the boys' and girls' distributions, then 3 more from the adults' distribution. Those draws resulted in the list I described in Section 3.2.1 and whose frequency distribution I plotted in Figure 3.1.

3.6   Hypothesis Testing and Frequency Distributions

A common question analysts must answer after calculating the frequency distribution of a data set is "what kind of distribution did this data come from?"

This question is the purview of hypothesis testing, a process in which the likelihood of a statement's being true is examined. Hypotheses that are found unlikely to be true and thus fail the test are labeled rejected. Hypothesis testing does not confirm that the hypothesis is true – other hypotheses could also explain the observed data just as well, if not better.

In the context of frequency distributions, hypothesis testing examines the probability that a particular distribution (usually a named one) could have produced the observed data. The basic process by which this is done is to compare the observed data to the expected frequency distribution predicted by the hypothesized distribution. If the two have markedly different shapes, the hypothesized distribution is rejected. The rest of this section covers the mathematics and statistics behind this process, but that's really the gist of it.

In this discussion, I will be working only with frequency distributions that have been discretized. For data sets measured with discrete scales the frequency distributions are already in this form, but for those measured with continuous scales a binning procedure like the one utilized to make a histogram must be used.

Previously, I referred to the measured frequency distribution simply as $N(q)$. In this discussion, I will instead refer to this quantity as $N_{\text{obs}}(q)$, meaning the observed number of elements as a function of the elements' values. In contrast, the number of elements that the hypothesized distribution predicts I will call $N_{\text{hyp}}(q)$ (some authors refer to this quantity as $N_{\text{ex}}(q)$, the number of elements the hypothesis tells you to expect).

How can I determine the expected frequency distribution predicted by the hypothesized distribution? The basic idea is to calculate, given n draws from the pdf $p(x)$ associated with that distribution, how many elements to expect to find in each bin of the frequency distribution – I can do this by working from the definition of a pdf and the number of elements n in the data set. Following the equation for $P(x_1, x_2)$ (Equation 3.11), the expectation for the population of a bin between x_1 and x_2 is given by:

$$N_{\text{hyp}} = n \int_{x_1}^{x_2} p(x)dx, \qquad (3.14)$$

which is the probability that a single element value would fall between x_1 and x_2, multiplied by the total number of elements.[11]

---

[11]For readers who are writing their own computer code, you can calculate N_{hyp} using a simple numerical integrator, like Simpson's rule. Simpson's rule states:

$$\int_{x_0}^{x_m} f(x)dx \approx \frac{h}{3}\Big(f(x_0) + 4f(x_1) + 2f(x_2) + 4f(x_3) + \ldots + f(x_m)\Big). \qquad (3.15)$$

where $x_i = x_0 + ih$ and $mh = x_m - x_0$. This is a weighted sum of the function evaluated at $m + 1$ points spaced evenly at a distance of h. The granularity of the points should be selected such that the function $f(x)$ is relatively smooth on that scale. In other words, if there are big jumps in f from point to point, the estimated value of the integral is likely to be far from the correct value. Run your code with several different h values to make sure the answer is converging.

Now that I can calculate the expected population of a given bin, how do I determine the bin ranges for which to calculate the hypothesized frequency distribution? There is a lot of freedom to choose whatever binning structure you like, but there are some mathematical restrictions associated with the simple hypothesis testing technique I'm describing here.[12] For this method to work properly, each bin should have N_{hyp} larger than 4 or 5 and the number of bins m should be larger than or equal to 4. So, choose a binning structure that satisfies the requirement for m and calculate N_{hyp} for each bin; does this configuration satisfy the requirement for N_{hyp}? If not, try a different binning configuration. When you have a binning structure that works, bin the observed elements using identical ranges. Then, with the bins indexed by j, the observed frequency distribution for a bin is given by $N_{\mathrm{obs},j}$ and the hypothesized frequency distribution for the same bin by $N_{\mathrm{hyp},j}$.

Finally, the two frequency distributions must be compared. One metric for doing this is the χ^2 statistic:

$$\chi^2 = \sum_{j=0}^{m-1} \frac{(N_{\mathrm{obs},j} - N_{\mathrm{hyp},j})^2}{N_{\mathrm{hyp},j}}. \tag{3.16}$$

The differences between the observed bin counts and the hypothesis predictions are known as the <u>residuals</u>; they appear squared in the sum's numerator. Hence, the χ^2 statistic is a sum of the residuals squared, divided by the hypothesis predictions. The term N_{hyp} appears in the denominator of the fraction for reasons related to the anticipated uncertainty in a counted quantity (see Taylor (1997) for more detail). If the predicted values N_{hyp} are close to observed N_{obs}, the residuals and the numerator of the sum will be small, minimizing χ^2.

Once χ^2 has been calculated, how can you determine if the hypothesis is rejected? First, look at the two frequency distributions plotted against each other; to your eye, are the data's major features captured by the hypothesized distribution? Next, examine the χ^2 value calculated from Equation 3.16. What size χ^2 should be expected to accompany a rejected hypothesis? The key variable for comparison is the number of bins, m. Note that χ^2 and m are both unitless counts, so they can be directly compared. If χ^2 is less than m, that is generally considered a plausible match, and if χ^2 is much larger than m, the hypothesis should be rejected.

In the previous paragraph, I slightly oversimplified the comparison of χ^2 to m. There is a third quantity that enters into determining whether the hypothesis should be rejected: the number of degrees of freedom in the hypothesized distribution. Essentially, the more parameters available to tune the hypothesized distribution, the lower χ^2 must be to not reject the hypothesis.

In Section 3.5, I discussed how I generated the height list from three normal distributions. Let's now imagine that I don't know how the height list was

---

[12]Technically, the most appropriate technique for a hypothesis test of a frequency distribution from a list measured with continuous scales is a Kolmogorov-Smirnov test.

generated. I now want to test the hypothesis that the elements corresponding to children's heights in the height list were drawn from a single Gaussian distribution. Since I'm just interested in the children's heights, I will remove the 3 largest heights from the data set; this produces a new data set with $n = 30$. From this data set, I calculate $\mu = 50.9$ in. and $\sigma_{n-1} = 1.7$ in.

I can use these parameters to specify a Gaussian-shaped pdf – that's my hypothesized distribution. I adjusted the bin width to satisfy the requirement that the expected number of elements in each bin be larger than 4 or so. I used $m = 6$, with 4 bins of equal width in the center of the frequency distribution and two of larger width at either end. The first and last bins include all of the probability associated with the tails of the Gaussian; in these bins, I fell slightly short of the minimum expected population of a bin. I can compare the hypothesized distribution with the data by performing the calculation defined in Equation 3.14 for each bin, then plotting N_{hyp} on the same axes as the frequency distribution. This is what I've done in Figure 3.4; the hypothesized distribution matches the data reasonably well, given that there are not many elements in this frequency distribution. The hypothesized distribution is larger than the data set's frequency distribution in the 49–50 in. bin, and smaller than the distribution in the 51–52 in. bin. For this hypothesis, $\chi^2 = 2.1$; when compared to the number of bins, this means that the hypothesis should not be rejected.

Of course, the children's heights were actually generated from two separate Gaussian pdf's, one for boys and one for girls. If that's the case, why wasn't my hypothesis that the heights came from a single Gaussian pdf rejected? Essentially, there aren't enough elements in the data set to distinguish between the two hypotheses! In one of the exercises at the end of this chapter, you will calculate χ^2 for the two pdf hypothesis, and demonstrate that neither hypothesis has a large enough χ^2 to be rejected.

You'll regularly see the Gaussian invoked as a model for a data set's frequency distribution, especially if the data set consists of a list of measurements of the same continuous quantity. This is generally done regardless of whether the data set is well-modeled by a normal distribution; for better or worse, the Gaussian is considered the standard for scatter around the mean. Many analysts will cram their data into a Gaussian distribution, throwing away outliers that don't fit the curve. This is a dangerous practice, as it is often the outliers that point to errors in the model.

3.7   Significance of Results

Finally, here is the discussion of "$X\sigma$" results that I promised you. You'll often hear people talking about an analytic finding as a 3σ result or something to that effect. These are statements regarding the likelihood that the

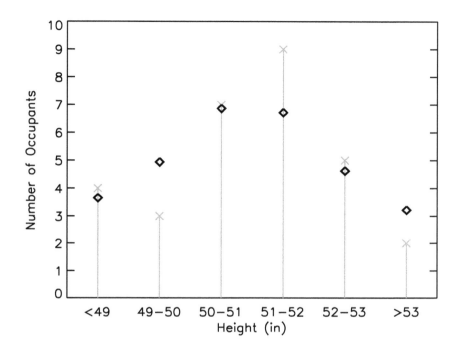

FIGURE 3.4
A comparative histogram of the height of the children in the height list (grey
crosses with vertical lines) with the Gaussian frequency distribution (black
diamonds).

result is significant; it's in other words, not due to random chance. Higher σ
results represent more confidence that the result is significant. 3σ results are
sometimes called "evidence" and 5σ results "discoveries."

The way these statistics are calculated is by comparing the actual data
to a model that doesn't include the new effect; this model is called the "null
hypothesis." In essence, analysts looking for the presence of a signal are trying
to disprove the null hypothesis that the signal isn't just caused by randomness,
and they do this with varying degrees of success. Note that you should always
examine how a null hypothesis is built, because dishonest analysts can choose
a null hypothesis that makes their results look more significant. As noted in
Chapter 1, this kind of behavior is counterproductive, because either someone
else will uncover your mistake or nature will bear out some other hypothesis.

There is a one-to-one mathematical relationship between the likelihood
that the result is true and the X in an $X\sigma$ result. Some common ones are 2σ
results that correspond to 95% confidence and 3σ results that correspond to
99.7% confidence. These equivalencies are calculated by integrating the area

under a normal distribution up to the specified number of σ. In statistics, the integral under the normal distribution is called the error function (erf), which is a function commonly built into software packages, and the specific relationship to likelihood is given by:

$$\text{likelihood fraction} = \text{erf}\left(\frac{z}{\sqrt{2}}\right). \qquad (3.17)$$

As usual, there are several different normalization conventions for the error function, and you'll want to test the error function in your programming language of choice with some simple examples before using it on your data.

3.8 Discarding Data from Normally Distributed Data Sets

By now, you should realize how careful you need to be if you choose to discard elements from your sample. However, there certainly are cases where the person taking the measurements makes a mistake – in these cases, throwing out that element would improve the analysis. Imagine trying to estimate the height of an average child from the height list, without knowing that there are three adults contaminating the sample. How can we know if removing these three elements is a reasonable step? Chauvenet's criterion (Taylor 1997) can help you make that decision, but only if you know the underlying distribution of elements should be Gaussian (this is very important).

The idea behind Chauvenet's criterion is to calculate for each element how many standard deviations it is located from the mean, then use the normal distribution to decide whether to expect a point that far from the mean, given the number of elements in the sample. Use this formula to calculate the number of standard deviations each point is from the mean:

$$\alpha_i = \frac{|x_i - \mu|}{\sigma_{n-1}}. \qquad (3.18)$$

Then calculate the expectation value for the number of points that should be located that far, or further, from the mean:

$$N_i = 2N_{\text{hyp}}, \qquad (3.19)$$

where N_{hyp} is calculated via Equation 3.14 for a bin with range (α_i, ∞) and the factor of 2 comes from the symmetry of the Gaussian (points on the left and right side of the mean). Traditional usage of Chauvenet's criterion implies that you can discard any point with $N_i < 0.5$, meaning that, on average, you would not expect to find a point that far from the mean. I would encourage you to choose a more stringent value.

Let's assume that the distribution of children's heights in the height list should be Gaussian.[13] Let's see if we can justify discarding the elements belonging to the adults. For this data set, we have seen that $\mu = 52.5$ in. and $\sigma_{n-1} = 5.2$ in. The three heights belonging to adults are located at 66 in. or higher, which is about 2.6σ from the mean. The value of N_i is 0.27, and there are 3 elements beyond at 66 in. or higher. This is an indication that something is wrong, and, if you were absolutely sure that the distribution should be Gaussian, you could discard these three elements.

At the risk of being redundant, I will emphasize that Chauvenet's criterion does not make the decision of whether or not to discard an element for you. First, you need to be absolutely sure that the data should be normally distributed. Second, you should have a plausible reason for how the elements entered the sample, whether it be observer error or something else. In the case of the height list, this rationale is that somehow adults' heights have entered into a sample that should solely be children's heights, perhaps due to a miscommunication between the person who collected the data and the person analyzing it.

3.9   Exercises

1. Imagine you had a list of the height of every adult in the United States. Would you use the mean or the median as a measure of the typical height? How strongly do you think the choice of technique would influence the analysis?

2. Find a list of the world's largest companies by annual revenue (for the purposes of this question, Wikipedia is an acceptable source). What do you first notice about the list? Plot the frequency distribution of this list in a histogram. What is the mean of this list? The median? The variance? The standard deviation?

3. Test the hypothesis that the list of the revenues of these companies is drawn from a Gaussian distribution.

4. Apply Chauvenet's criterion to the revenue list – can you justify throwing out some of the highest revenue companies? Is this a good or a bad idea?

5. Test the hypothesis that the children's heights in the height list are drawn from two Gaussians, as described in the footnote in Section 3.2.1, both with $\sigma = 1.5$ in. Can this hypothesis be rejected?

---

[13]This assumption is, in fact, incorrect, since the distribution for the children is actually the sum of 2 Gaussians, one for boys and one for girls.

6. Test the hypothesis that the age list is drawn from a single Gaussian pdf. Can this hypothesis be rejected?

4

Uncertainty and Error

4.1 Introduction

Part of being a responsible analyst is communicating how certain you are of your findings. Analysts use uncertainty and error analysis to accomplish this task. Why is this an important topic? Put bluntly, people make better decisions when a finding's certainty is properly communicated. It's impossible to predict the types of decisions that people will base on your analyses, but there can be money or, more seriously, lives at stake. Good analysis with appropriate uncertainty communication results in better outcomes.

For example, imagine a decision maker considering two conflicting pieces of information. How can she weigh their relative merits? If it turns out that one of the pieces of information is an extremely uncertain result, that simplifies the decision maker's job because she knows that she can put more trust into the more certain result, and the decision she is making is more likely to lead to a preferred outcome.

From this example, I'm sure you can see the perverse incentive facing analysts: if you state a conclusion with false confidence and optimistically small uncertainty, you are more likely to influence the decision maker. However, from the selfish perspective of protecting your reputation as an analyst, it is self-defeating to pass off borderline results as being certain by avoiding or downplaying error analysis. In these cases, you're bound to be wrong more often than not, and then your credibility will have suffered harm.

Uncertainty isn't limited to analytic findings. Most (but not all) measurements have associated uncertainty, as do the intermediate quantities in an analysis. Later parts of this chapter will describe how the error in your measurements propagates to the error in your findings.

4.2 Definitions, Descriptions, and Discussion

I'll start the discussion of error and uncertainty by introducing some of the technical terms and concepts used by analysts. Later, I'll move on to a discussion of the mathematics.

In this book, I will use the terms <u>error</u> and <u>uncertainty</u> interchangeably; they will both refer to the level of confidence an analyst has in a measurement or a finding.[1] In other words, error and uncertainty describe the difference between the analyst's knowledge and the true value of a behavior of interest.

There are different kinds of error and, unfortunately, the definitions of these error subcategories aren't standardized across authors.[2] Furthermore, even when authors take great care to differentiate between error categories, there are still a large number of grey areas, leading to confusion in classifying and communicating uncertainty. My advice is to not treat the categories that follow as mutually exclusive. Instead, imagine situations in your own analyses and the examples from Chapter 3 where these types of uncertainties could occur and the effects they could have on your conclusions.

4.2.1   Illegitimate Error

When most people hear the word "error" they think of mistakes (a topic discussed in Section 1.4.2), but this meaning is quite different from the technical definition analysts use. Mistakes are instead characterized as <u>illegitimate error</u>; examples include inputting a wrong number into a spreadsheet or making a mistake in a piece of computer code that affects your results. The effects and the likelihood of illegitimate error are extremely difficult to estimate or predict. Your best hope to minimize illegitimate error is to thoroughly inspect your derivations, calculations, and code. Confirm that your calculations produce the expected results when you set parameters to extreme values or to zero (this is called checking the "limiting cases"). In any event, the rest of this discussion does not pertain to illegitimate errors.

4.2.2   Where Does Error Come From?

If error analysis does not mean checking your work for mistakes, what does it mean? Error is related to the act of measurement, and error analysis is primarily derived from the measurement of physical quantities. However, for the reasons I discussed in Section 4.1, error analysis is important for all analytic

---

[1]This is far from a universally accepted definition. It's the one I use because it's common in the physical sciences (e.g., Taylor (1997)), the field where I was trained. However, other fields use the terms differently. For example, many analysts define error identically with accuracy and uncertainty identically with precision. (I will discuss how I use accuracy and precision shortly.) When communicating with other analysts, be aware that these words don't have standardized meanings.

[2]Some analysts draw a distinction between the terms uncertainty and risk. For these analysts, risk is error that can be mathematically modeled – it's a "known unknown." Uncertainty is error that is potentially much larger than the modeled uncertainty; it occurs because of a lack of understanding of the measured behavior. In other words, uncertainty is composed of the "unknown unknowns" (see Section 4.6). Personally, I don't use these terms in this way because of the long history in scientific measurement behind the word "uncertainty," but the concept underlying the distinction is sound, and some analysts prefer it (Knight 1921; Silver 2012).

conclusions, even those that aren't directly related to the physical world and don't involve a measurement apparatus.

Measurement entails putting a (usually) numerical quantity to some real world behavior of interest. In most cases, you can think of measurement as an attempt to determine the unknown true value of a quantity. For example, imagine I'm interested in the weight of a bowling ball. The bowling ball has some true weight that I don't know, so I use an apparatus (such as an analog bathroom scale) to measure that weight. However, because the scale isn't perfectly constructed, its display is hard to read, or because the ball is hard to keep stationary on the scale, the measurement won't result in the exact true value.[3] Instead, it will return some other value influenced by all of the effects I mentioned, each a source of measurement error.

A simple way to demonstrate measurement error is to take repeated measurement of the same behavior of interest – the measurements will likely not return identical values. For example, each time you measure the weight of the bowling ball with an analog scale, your measurements will differ slightly due to variations in the apparatus setup or in how you read the scale.[4] To better understand the variability of repeated measurements, you can think of these measurements as being drawn from a pdf influenced by both the characteristics of the behavior of interest and the technique and apparatus used to measure it. This pdf is known as the parent distribution or, alternatively, as the limiting distribution. If you are using a high-quality measurement apparatus, the parent distribution will be tightly centered around the true value (though variability in the behavior of interest can widen the parent distribution). If your apparatus is of poor quality, the true value will be skewed from the center of the parent distribution and the measured values will be wildly scattered.

When a measurement is taken, the value is drawn from the parent distribution; the frequency distribution of repeated measurements of the same quantity is known as the sample distribution (Bevington and Robinson 2003). Note that in most cases, it is impossible to determine the parent distribution exactly because there is no way to know the true value, but you can approximate the parent distribution by increasing the number of samples drawn. In the limit of infinite samples, the shape of the sample distribution will match the parent distribution. In Section 3.6, I discussed how to test whether a hypothesized distribution could produce an observed frequency distribution; in the language of this chapter, what I was testing was whether a particular parent distribution could produce the sample distribution.

In the case of the height list, defined in Section 3.2.1, I have already stated that the parent distribution is a mix of three Gaussians: one for the boys,

---

[3]Quantum mechanical behavior requires a special model for measurement, but unless you're a physicist or someone working with those behaviors it's probably best not to worry about it.

[4]Not all quantities have uncertainty in repeated measurements – I can count the number of desks in my office and get 1 each and every time.

one for the girls, and one for the adults. It is only because that particular data set is not drawn from real measurements that I can actually tell you the parent distribution! I have also already plotted the sample distribution; it's the histogram you remember from Figure 3.1. To compare the sample to the parent distribution, I plotted them on the same axes (Figure 4.1). Technically, I've plotted the sum of the parent distributions for the boys, girls, and adults scaled by the total number of samples.[5] Note how the sample distribution and the scaled parent distribution have slightly different shapes; the sample distribution falls short of the parent distribution on the high side of the larger peak near 53 in. and overshoots the parent distribution at 51 in. Additionally, the parent distribution at larger heights doesn't quite reach 1 for any particular height, but the three draws from the adult parent distribution have to appear somewhere (in other words, the binned sample count is discretized to positive integers). If the number of samples were increased, the shapes of the sample and parent distributions would converge.

It is impossible to make a plot similar to Figure 4.1 for the age list, since it is a real data set for which we don't know the parent distribution.

In short, when a measurement is made, uncertainty causes the sample drawn from the parent distribution to differ from the true value of the behavior of interest. This uncertainty can be split into two components: random and systematic error.

4.2.3 Random Error

Random errors, also known as noise, result from natural variability in the analytical apparatus and in the system being studied. Random errors are the errors most often quoted in analytic conclusions, usually in the form of error bars. Taylor (1997) describes a straightforward rule for distinguishing between random and systematic errors: uncertainties that can be revealed by repeating the measurements are random, and those that can't are systematic. Keep in mind that characterizing an effect as random error does not mean that it is impossible to understand its origin – if you looked deep enough into what actually caused different values to be measured in repeated measurements, you may be able to find an explanation.

I will delve into the mathematics of random errors in Section 4.4. As you carry out analyses, don't forget that the reason authors present the theory and mathematics of random errors is because they are easy to understand, model, and use. However, this does not mean random errors are most important type of error! In fact, many analysts believe that more analytical progress happens as a result of the reduction of systematic errors rather than random ones.

---

[5] In order to make a fair comparison between the data and the parent distribution I should actually be plotting binned versions of the three parent distributions, as I did in Figure 3.4, but here I want to show how the smooth shape of the limiting distribution is echoed in the frequency distribution.

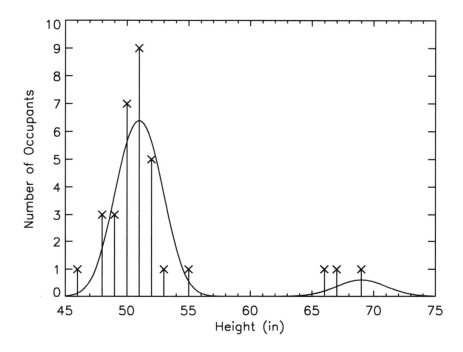

FIGURE 4.1

A histogram of the height list, with the expected number of measurements derived from the three parent distributions overplotted. The leftmost peak is actually two Gaussians, one centered at 50 in. (boys) and the other at 52 in. (girls). The rightmost peak centered at 69 in. corresponds to the adults. The curve was derived from the three parent distributions by scaling each by the number of measurements: 15 from each of the two student genders, and 3 from the adults.

4.2.4 Systematic Error

Systematic error causes a set of measurements not to be centered around the true value, but instead to be biased in one direction or another away from it. In other words, systematic error displaces the mean of the parent distribution from the true value. Systematic errors can't be minimized or eliminated by repeated measurements, only by improving the measurement technique or apparatus.

The history of science is littered with examples of unaccounted for systematic errors; for example, later experiments showed that the value of the charge of the electron measured by Robert Millikan in 1909 in the famous "oil-drop" experiment was too low by approximately 1%, a factor much larger than the uncertainty he estimated in the measured value. His result was biased because he used an incorrect value for the viscosity of air, a parameter totally unrelated to the quantity he was measuring (Feynman 1997). Repeating his measurement over and over would not have allowed Millikan to reduce this systematic error. Though Millikan was (deservedly) awarded the Nobel prize for this work, the original measurement is an example of a dangerous situation: a measured quantity whose distance from the true value is far larger than the quoted uncertainty. Even the best analysts will make these kind of mistakes, not because they are careless, but because analysis is hard.

As a simpler example of systematic error, consider a stopwatch calibrated to run at standard room temperature, but being used at an ambient temperature well below freezing. The internal mechanism of the stopwatch may be slowed by the cold conditions, resulting in recorded times that are consistently shorter than an accurate watch would measure. Note again that repeated measurements will not reduce this type of bias.

Systematic errors are very difficult to identify and isolate. The best practice for doing so is to consider every step of the measurement process, and brainstorm every possible way something could go wrong. Once you have identified a systematic error, try to estimate the amount of bias associated with it. This estimation is rarely straightforward, and sometimes impossible. Whether you can determine an estimate or not, your conclusions should contain proper caveats documenting the limitations resulting from systematic errors.

Not all systematic errors can be eliminated. Millikan would have used a better value for the viscosity of air had he known it, but it may not be a simple matter to find a stopwatch that is insensitive to cold. Therefore, as a general rule, try to design your analyses such that the random error in the measurement is much larger than your estimate of the systematic error. If you follow this principle, you will generally avoid the dangerous situation of the true value of your measurement being outside the range of your uncertainty estimates.

4.2.5   Aleatory and Epistemic Uncertainty

Many fields find it useful to separate uncertainty into natural variability in the system (aleatory) and uncertainty due to lack of knowledge (epistemic) (Bedford and Cooke 2001). Aleatory uncertainty is essentially a form of irreducible random error, whereas epistemic uncertainty can be reduced by studying the system further and developing a more detailed model. Of course, even these definitions are arbitrary, as the only limits on how detailed a model you can build are quantum mechanics and chaos theory. One experiment's aleatory uncertainty is a more detailed experiment's epistemic uncertainty. However, judicious estimation of epistemic uncertainty is useful, in that it tells you how much better your measurement could get, should you be willing to invest the effort to improve the model or measuring apparatus. Of course, similar to random error, it is generally much easier to estimate the aleatory uncertainty than the epistemic variety.

4.2.6 Precision and Accuracy

The terms precision and accuracy are related to uncertainty, but they are often used outside of technical discussions as well. When used as terms of art, they have specific and distinct meanings. Precision refers to the degree to which a measurement has constrained the value of the behavior of interest. It does not depend on the relationship between the measurement and the true value of that quantity; it is only a characteristic of the measurement. Some authors distinguish between absolute precision, which is simply the size of the constraint the measurement has placed on the true value, and relative precision, which is the size of the constraint divided by the best estimate (Bevington and Robinson 2003). In either case, if two experiments measured the same quantity, the one with the smaller uncertainty could be said to be more precise, though that doesn't imply that it is more correct. For that to be true, it would have to be at least as accurate as the other measurement – accuracy relates to whether a measurement actually corresponds to the true value of the measured quantity. For example, if you have a large systematic error in your experiment, the accuracy of your measurement is likely to be poor, even though the precision could be quite good. A perfectly accurate experiment has no bias or systematic error, though there still could be random error in the result (Mandel 1964).

4.3 Quantification of Uncertainty

Quantifying the error around a conclusion is the best way to communicate the analyst's understanding of its certainty. Surprisingly, not all fields take a quantitative approach to communicating uncertainty. For example, in the field of intelligence analysis, qualitative conclusions are communicated with linguistic descriptors of their certainty. An intelligence analysis might state "There is a serious possibility of event X occurring." However, studies have shown that qualifications such as "serious" and "likely" do not have universal agreed-upon meanings. Kent (1964) has shown that analysts working *on the same study* did not agree on the meaning of these terms; when they agreed that the occurrence of an event was a "serious possibility," the likelihoods stated by the analysts ranged from 20% to 80%. The analysts didn't agree on what they were agreeing on! Surely this makes for meaningless conclusions. In contrast, quantitative measures of uncertainty have definitive meanings; they clarify the discussion, instead of muddying the waters. Let's take a look at two different ways to quantify uncertainty.

4.3.1 Uncertainty Distributions

An <u>uncertainty distribution</u> is a general quantitative method for characterizing uncertainty related to the pdfs $p(x)$ described in Section 3.5. An uncertainty distribution is a special case of a pdf in which the distribution characterizes the likelihood that the true value of the measured quantity is x. These distributions can be constructed for both discrete and continuous x. As with all pdfs for continuous x, you'll need to integrate over $p(x)$ to find the probability that x is found between x_1 and x_2 (see Equation 3.11). In the simple case where the uncertainty distribution is a Gaussian, the most likely value for x is found at the peak of the curve, which also happens to be the mean of the curve. In the case of an arbitrary distribution this rule doesn't hold, and it is best to include the most likely value of the parameter, the mean value, and some measure of the range of likely values in the communication of uncertainty.

There are some measurements that can only be clearly communicated as an uncertainty distribution. For example, consider a computer hard drive reader that measures a bit b with a misread rate of 1 in 10^4. If the reader measures the bit as '1', the uncertainty distribution $p(b)$ is given by $p(1) = 0.9999$ and $p(0) = 0.0001$.[6]

4.3.2 Error Bars

In many analyses, uncertainty is presented in terms of a single number as shorthand for a fully-described uncertainty distribution. Generally, translating a distribution into a single number is a lossy process, meaning that much less information is contained in the single number than in the full distribution. This isn't always true, since a Gaussian distribution can be fully characterized by just two numbers (the standard deviation with the mean as the best estimate), but as I discussed in Chapter 3 you shouldn't assume distributions are Gaussian without evidence.

When uncertainty is reduced to a single number, it is usually presented in the following form: $x \pm \delta_x$. The first term (x) corresponds to the best estimate and the second term (δ_x) corresponds to the <u>error bar</u> associated with that quantity. The combination of best estimate and error bar is a simple method of presentation, but the term δ_x, which characterizes the scale of the uncertainty, doesn't have a consistent quantitative meaning. In some contexts, it is the uncertainty in the mean of a series of repeated measurements, σ_μ, described in Section 4.5. In other cases, it is the half width of a range that contains 68% of the probability of including the true value. When the distribution is Gaussian, these two definitions are equivalent. However, in many other cases, it is the half width of the 95% probability range, equivalent to $2\sigma_\mu$ for a Gaussian

---

[6]Technically, this example also assumes that the 0 and 1 bits were equally likely before the measurement was made. If this doesn't make sense to you now, check back after you've read Section 8.4 on Bayesian analysis.

distribution. Most importantly, even when the distribution is decidedly not Gaussian and has some completely arbitrary shape, the uncertainty is often presented in the same way.

In summary, you should be careful when presenting uncertainty as an error bar or when interpreting someone else's presentation, because there is no consistent standard for the meaning of δ_x. If you choose to present uncertainty in this format, be sure to also communicate the method by which your error bar was derived.

Two common variations of error bars are worth being aware of. One variation occurs when the error bar is asymmetric, meaning it's larger on the positive than the negative side (or vice versa). This shouldn't happen when error bars are used to describe Gaussian uncertainty distributions, since they are symmetric, but as I've stated analysts use error bars with many different meanings. Asymmetric error bars are generally written as $x^{+\delta_{x1}}_{-\delta_{x2}}$. For example, a length measured to be 1.0 cm with uncertainty bounds given by 0.8 cm and 1.5 cm would be written as $1.0^{+0.5}_{-0.2}$ cm. A second variation occurs when an analyst wants to present random and systematic errors separately. In this case, the quantity is written as $x \pm \delta_{x,\mathrm{ran}} \pm \delta_{x,\mathrm{sys}}$.

4.4 Propagation of Uncertainty

In this section I will discuss calculating the uncertainty of a quantity that is itself a function of quantities with uncertainty. This topic is called the propagation of uncertainty, because the known error in constituent quantities is carried through mathematical processes to find the uncertainty in the final product. This section draws heavily on Chapter 3.7 of Taylor (1997), which contains a terrific discussion of why the rules for propagation take on the forms I'm about to describe.

There are two limiting cases for the propagation of uncertainty that are distinguished by the linkages between the known errors. If the errors in the constituent quantities are random and independent, the uncertainties do not add linearly, because they sometimes work to cancel each other out. On the other hand, if the errors in the constituent quantities are correlated, they are more likely than not to reinforce each other. Fully correlated errors result in the largest errors in the final product.

As examples of these two cases, imagine the following situation. You own a company that manufactures a piece of hardware that contains a widget and a sprocket, each of which is made of yet smaller parts. You are measuring the amount of time it takes one of your employees to assemble a widget and a sprocket from their components. The time to construct the widget is measured, and then the time to construct the sprocket measured separately, then the times are added to create a total time:[7]

$$t_{\text{total}} = t_{\text{widget}} + t_{\text{sprocket}}. \qquad (4.1)$$

Consider the differences between the two measurement apparatuses that follow. In the first approach, time is measured by clicking an electronic stopwatch when each part is complete. Assuming that the dominant source of error is the clicking of the button by hand, and that the observer's clicking ability is accurate but not very precise (meaning the parent distribution is centered around the true value), the error in the total time can be treated as the sum of two independent errors. In the second situation, imagine a device that automatically records the time when a part is complete, but the device is sensitive to the temperature, like the watch we described in Section 4.2.4. In this situation, the error in the two times is systematic and correlated, so the uncertainties are more likely than not to reinforce each other. With this second apparatus, you should use the worst-case formulae.

4.4.1   Worst-Case Errors

For the next few sections, I will describe how to determine the error in quantity z, given by δz, when z is a function of known quantities $a \pm \delta a$ and $b \pm \delta b$. In other words, if $z = a + b$, use the addition rule to determine δz from δa and δb. If $z = ab$, use the multiplication rule. These rules apply to error bars of the 1-σ variety.

If the errors in the known quantities a and b are correlated, or if it is unclear whether they are or not, the following formulas give the worst case for the propagation of errors into δz:

$$\delta z \quad = \delta a + \delta b \quad \text{Addition and Subtraction} \qquad (4.2)$$

$$\frac{\delta z}{|z|} \quad = \frac{\delta a}{|a|} + \frac{\delta b}{|b|} \quad \text{Multiplication and Division} \qquad (4.3)$$

You can imagine how these worst cases are derived; clearly in the case of perfect error correlation and $z = a + b$, the uncertainties δa and δb should simply add to produce δz. Anticorrelation and subtraction result in the same combination.

4.4.2   Random and Independent Errors

For independent random errors, the uncertainties add in quadrature, meaning as the square root of the sum of the squares, rather than linearly.

$$\delta z \quad = \sqrt{\delta a^2 + \delta b^2} \quad \text{Addition and Subtraction} \qquad (4.4)$$

$$\frac{\delta z}{|z|} \quad = \sqrt{\left(\frac{\delta a}{|a|}\right)^2 + \left(\frac{\delta b}{|b|}\right)^2} \quad \text{Multiplication and Division} \qquad (4.5)$$

---

[7]The exact form of this equation doesn't matter for this example, just the fact that there is one quantity that depends on two measurements.

Test out these formulae and convince yourself that the random and independent errors work out to be equal to or smaller than the worst-case errors.

4.4.3 Other Common Situations

Two frequently occurring situations involve the measured quantity $a \pm \delta a$ operated on by an exact number to produce $z \pm \delta z$. When a is multiplied by the exact number E to produce z ($z = Ea$), the new error is predictably given by:

$$\delta z = |E|\delta a \quad \text{Exact multiplication} \tag{4.6}$$

When a is raised to the exact power E to produce z ($z = a^E$), the new uncertainty is given by:

$$\frac{\delta z}{|z|} = |E|\frac{\delta a}{|a|} \quad \text{Exact power law} \tag{4.7}$$

There are no separate worst and independent errors in these situations because z only depends on one variable, so there is no second variable for it to be correlated with.

4.4.4 The General Case

The power law equation has likely clued you in to the general form of these error propagation laws. For z calculated as a function of several variables, $f(a, b, \ldots)$, the worst-case version of the error formula is:

$$\delta z = \left|\frac{df}{da}\right|\delta a + \left|\frac{df}{db}\right|\delta b + \ldots \tag{4.8}$$

where df/da corresponds to the derivative of the function f with respect to the variable a

For independent and random errors, the errors are summed in quadrature:

$$\delta z = \sqrt{\left(\left|\frac{df}{da}\right|\delta a\right)^2 + \left(\left|\frac{df}{db}\right|\delta b\right)^2 + \ldots} \tag{4.9}$$

4.4.5 Non-Uniform Uncertainty in Measurements

When there is no uncertainty in a data set, or each element has the same uncertainty, calculating the mean of the sample proceeds as described in Section 3.3.1. In the more general case where the measurements are given by $x_i \pm \delta x_i$ (1-σ Gaussian errors), the mean should be constructed with a weighted sum (Bevington and Robinson 2003):

$$\mu = \frac{\sum_{i=0}^{n-1}\left(x_i/\delta x_i^2\right)}{\sum_{i=0}^{n-1}\left(1/\delta x_i^2\right)}, \tag{4.10}$$

where the measurements with the smallest uncertainty enter with larger weights.

4.5 Uncertainty in the Mean

Let's look more closely at the case where the uncertainty is dominated by random error. If the mean of a sample with identical error in each element[8] is calculated to be μ, how close is that quantity to the mean of the parent distribution? There is approximately a 68% chance that the sample mean will fall within σ_μ, the uncertainty in the mean (also known as the standard error of the mean), when the parent distribution is Gaussian:[9]

$$\sigma_\mu = f \frac{\sigma_{n-1}}{\sqrt{n}}, \tag{4.11}$$

where f is a correction factor for finite sample size, discussed below. For now, you can think of f as a factor close to 1. Note the dependence of σ_μ on the number of measurements n, implying (reasonably) that we can increase our confidence in the fidelity of our measured mean by increasing the number of samples. Of course this is true, since as the number of samples increases, the sampled distribution approaches the parent distribution! I referred to this effect earlier in the chapter, when I discussed how random error could be decreased by repeated measurement.

However, there are two relevant caveats to the reduction in σ_μ through repeated measurements. First, the number of samples enters as the inverse square root. This means that if I want to reduce the uncertainty σ_μ by a factor of 100, I need to take 10^4 more measurements. This quickly gets tedious. Secondly, this analysis assumes that the uncertainty is dominated by random errors. As σ_μ from random errors decreases, the relative influence of systematic error increases. In other words, even if you drive the random σ_μ to 0, systematic error will still be present in your findings.

It's worth looking at the height list from Chapter 3 to see how the uncertainty in the mean works in practice. If I assume the measurement uncertainty in the list elements is very small, I can calculate the average height of a child in the classroom (the mean of the sample) with essentially zero error. If I assume the uncertainty in each measurement is ± 1 in. (1-σ Gaussian error bars), I could use the rules presented in Section 4.4 to calculate the error in

---

[8]Uncertainty in the weighted mean is discussed in Taylor (1997) and Bevington and Robinson (2003).

[9]As I've discussed, that's not always a safe assumption, but you can check the assumption by examining whether the frequency distribution of the sample has a Gaussian shape (see Section 3.6). Calculating the uncertainty in the mean for an arbitrary parent distribution is an advanced topic (see Bevington and Robinson (2003)).

the height total, and divide that by the exact number n to find the uncertainty in the sample average. However, neither of these quantities represents my uncertainty in the mean of the larger population, because neither has been connected to the parent distribution from which the data are sampled.

Think about this question: what exactly does the parent distribution from which the samples were drawn represent? It isn't the distribution of the classroom kids' heights, because I know that sample distribution exactly. In this case, the parent distribution could be something like the distribution of the heights of the kids living in that neighborhood, of which some fraction are members of the class. Or maybe it's something like the distribution of the heights of kids of that age across the country. There's no right answer here, since I made the distribution up. The important point is that the standard error in the mean tells you how well you've pinned down the mean of that parent distribution, not the mean of the sample. In this case, from measurements of the 30 kids (and excluding the three adults), I've determined $\mu = 50.9 \pm 0.3$ in., where I'm stating the 1-σ random uncertainty in the mean. This is a pretty good estimate, since the actual parent distribution mean is 51 in.[10]

There are a couple of potential complications here. Let's pretend that this is a real data set where the parent distribution is unknown, and I'm interested in using the heights of the kids in this class to estimate the mean height of the kids across the country. There's a lot of room for systematic error here, specifically in the form of bias. What if the children in this neighborhood were malnourished and their growth was stunted? Clearly the estimate of μ I derive from my sample will be biased lower than the parent distribution's mean. Most importantly, that systematic error *will not* be reflected in σ_μ, because Equation 4.11 only accounts for random error! How could you estimate the systematic error? Well, you could start by doing some research on how much malnourishment stunts a child's growth, and adding that as a separate uncertainty. Of course, it would be much easier to use a representative sample of children's heights across the country, but you may have to make do with the data you actually have.

Another example of potential systematic error in this case is from sample contamination. It is possible through some carelessness that the adults in the room are included in a sample that is supposed to be exclusively children, and that information is not passed along to you, the analyst (you could also characterize this as illegitimate error). The contamination will bias μ and drive its associated random error upwards to $\mu = 52.5 \pm 0.9$ in. However, the calculated uncertainty in the mean is problematic because the parent distribution is no longer Gaussian, violating one of the initial assumptions for using Equation 4.11.

As noted in Section 4.2.4, the key is to drive your systematic errors down

---

[10]Remember, this parent distribution consists of two Gaussians, one with a mean of 50 in and a second with a mean of 52 in. I've assumed that these two Gaussians merge to form a distribution that itself is roughly Gaussian.

to the point that they are dominated by, or at least significantly smaller than, your random errors. This is not always easy to do.

Standard Error of Proportion

Data sets measured with categorical and ordinal scales have an analogous statistic to the uncertainty in the mean called the standard error of proportion (also known as the margin of error, which you may be familiar with from political polls). Let's say you've calculated the percentage of elements in a data set that fall into one category or ordinal bin (such as an expressed preference for candidate A). Call that percentage p. If your data set is a random sample of a larger population, as it often is in polling, how much could p differ from the percentage of that category or ordinal bin in the larger population?

The standard error of proportion, defined as the 1-σ uncertainty in a proportion, is given by:

$$\sigma_p = f\sqrt{\frac{p(1-p)}{n}}, \tag{4.12}$$

where f is the same correction for finite sample size in the uncertainty in the mean. Notice that in the limit of a large population, the standard error of proportion decreases as the square root of the number of measurements, just like the standard error of the mean does.

The standard error of proportion is the relevant measure of uncertainty in a variety of scenarios. For example, if you perform a medical test on a random sample of the U.S. population and some fraction test positive, how well has the fraction of infected people in the entire U.S. population been determined? If you examine a random sample of tax returns from businesses for 2012 and some fraction of those returns show an annual profit, how well have you determined the percentage of profitable businesses in the United States for the year? If you flip a coin 1,000 times and it comes up heads 450 times, how sure are you that the coin is not fair?

Equation 4.12 is derived using approximation techniques and is therefore only valid when certain conditions are met. In most situations, it can be used safely when $np > 5$ and $n(1-p) > 5$ (other authors present similar but slightly different conditions for this formula's use). For applications that violate these conditions and for those where high precision in the uncertainty estimates is required, see the discussion in Brown et al. (2001). This is particularly important when p is close to 0 or 1.

Now, for the discussion of the f in Equation 4.11 that I promised earlier. Sometimes, a data set's elements are samples drawn from a population that is larger than the sample, but not infinitely large. For example, consider a bucket of similar sprockets from which you draw a handful and measure their diameters. When the sample is a significant fraction of the larger population, a correction is necessary to reduce the uncertainty in the mean compared to a situation where the population is infinitely large. The correction for finite population size, f, is given by:

$$f = \sqrt{\frac{L-n}{L-1}}, \qquad (4.13)$$

where L is the size of the larger population that the sample is drawn from and n is the sample size. For this formula to be appropriate, it's important that the samples are chosen from the larger population without replacement, meaning that there is no chance a single sample could be drawn twice. Notice that in the limit where the sample size n is much smaller than population size L the correction term is approximately 1; this is the infinite population limit. Conversely, as n approaches L, the correction term approaches zero; this means that the closer the sample gets to containing the entire population, the smaller the uncertainty in the estimate of the mean. Note that f is always less than 1, meaning that taking into account the finite nature of the population reduces the estimate of the uncertainty in the mean.

For repeated measurements of the same quantity, such as measuring the length of a piece of string with a ruler, treat the larger population size as infinite, because you could repeatedly measure that quantity until the end of time. It's those infinite potential measurements that you are sampling from.

4.6 Why Is This So Unsettling?

Well, it should be.

Let's imagine you've constructed a model for the daily movement of a new stock which has a current price of $1000. You've monitored the price over the past 1000 business days, and found that the distribution of daily moves is well fit by a Gaussian with $\mu = \$0.10$ $\sigma = \$1$; let's call this the assumed model. Over the long run, a stock that followed the assumed model would be a profitable bet, since on average its price should rise by $1 every 10 days. However, what if the price's daily movement, x, was actually drawn from this pdf:

$$p(x) = 0.9999 \times \frac{1}{\$\sqrt{2\pi}} e^{-\frac{(x-\$0.10)^2}{\$2}} + 0.0001 \times \frac{1}{\$} \times \delta_D(x = -\$900), \qquad (4.14)$$

where δ_D is the Dirac delta function, which integrates to 1, and follows this behavior:

$$\delta_D(x = c) = \begin{cases} \infty, & \text{for } x = c \\ 0, & \text{for } x \neq c. \end{cases} \qquad (4.15)$$

In other words, the pdf for the stock's daily movement looks like a Gaussian

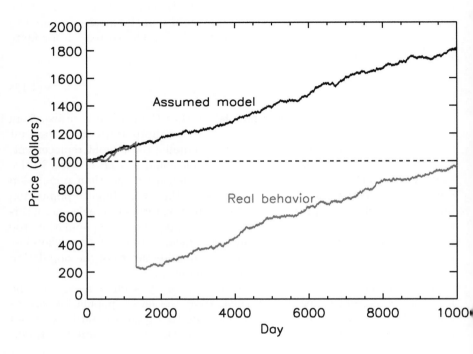

FIGURE 4.2
The "Assumed model" curve is a simple Gaussian model, which rises slowly
in price. The "Real behavior," which contains a small likelihood of a big drop,
has a very different outcome.

centered at a positive value just above 0, with a spike at a larger negative
value. The units of inverse dollars are present so the pdf will integrate to a
unitless probability. This means that there is a very small chance (0.01% of the
time, or one out of every 10,000 days) that the stock will lose $900. Running
a quick 10,000 day simulation, how does this play out?

Admittedly, this isn't a realistic model for how a stock would behave, but
let's examine the results anyway (Figure 4.2). The addition of an event that
occurs with only a 1 in 10,000 probability dominates the behavior of the stock.
The presence of this feature in the distribution completely determines whether
or not the stock is a good long-term investment. Using a simple Gaussian pdf,
you would have predicted a drop in the price of that magnitude to occur with
a likelihood too small for my computer to calculate (i.e., a 900σ event). And
from researching the stock over a fairly long period of time (3 yrs), you have
no idea that this type of behavior is possible.

This behavior is a general problem because analysts tend to have the most
confidence in the part of the model that occurs most frequently. In a normal

distribution, this is the area around the mean. The analyst has the least amount of knowledge about the tails of the distribution, which are generally fit with a model whose shape is determined by the area around the most frequent events. This is a dangerous (and difficult to avoid) way of doing business! The tails of the distribution, which can dominate the results (especially in the long term), are the least well-determined part of the distribution.

In practice, in complex systems events that our models tell us should be very rare happen more often than expected (Taleb 2007). When these events occur, they are signals that the model was never an accurate depiction of the system in the first place. In economics, the uncertainty associated with the failure of a model to predict major events is known as Knightian uncertainty (Knight 1921).

The bottom line is simple: don't assume your models and errors are Gaussian unless you have a very good reason to do so. Another good rule of thumb is to not make predictions beyond the timescale over which you have observed the system in question.

4.7 Exercises

1. Using the general case formula for error propagation, calculate the equation for δz for the functional relationship $z = (a^2/b) - \sin c$ when there are errors in a, b, and c. Compare and contrast the error propagation in this case for independent and worst-case errors. This question requires knowledge of derivatives.

2. Suppose each element in the height list has an associated 1-σ uncertainty, normally distributed, of 3 in. for all the elements less than 60 in., and 5 in. for all larger elements. For each element, calculate the "Absurd Height Index," A, defined by $A_i = x_i^2 + \exp(x_i/1 \text{ in})$, along with the elements' associated uncertainty. For this calculation, why doesn't it matter if the uncertainty is systematic or random? This question also requires knowledge of derivatives.

3. (Advanced) Using random draws from the three parent distributions for the height list confirm the sample distribution approaches the three parent distributions in the limit of a large number of samples.

5

1-Dimensional Data Sets

5.1 Introduction

A <u>1-dimensional data set</u> is a structure in which each of the constituent elements is measured and indexed along a single dimension. The 1-dimensional data set, also known as a 1-dimensional array, is the archetypical subject of many foundational quantitative analytic techniques. For example, it is the simplest type of data set on which to perform frequency analysis, a powerful investigatory tool. Useful cleaning and filtering techniques can also be applied to rid 1-dimensional data of misleading systematic errors. As usual, I'll discuss situations where these techniques are appropriate, as well as some common ways they are misused.

The formalism of 1-dimensional data sets is similar to that of lists. Let's define a structure of n elements ψ_i starting with the zeroth element, along with a structure x_i of values marking the locations where the elements ψ_i are measured. The last element in the sequence, ψ_{n-1}, is measured at the location x_{n-1}. The values of x generally correspond to a physical dimension, like time or distance, and are monotonically increasing or decreasing.[1]

Some examples of 1-dimensional data sets are:

- The temperature at the airport, measured throughout the day,

- The height of a wire strung between two telephone poles, measured as a function of position, and

- The price of a stock measured at market close for a series of days.

These examples illustrate the structural difference between lists and 1-dimensional data sets: because the elements ψ_i are indexed to x_i, the order of the elements in ψ contains information. If you swapped the airport temperatures measured at two different times, the data set would be altered. This is not true for the height list from Section 3.2.1, where a change in the element order does not affect the analytic conclusions.

The lessons I discussed about the process of analysis in Section 1.3 also apply to 1-dimensional data sets. In particular, my recommendation to begin by

---

[1] There are occasions where x_i can simply be the number of the observation – performing temporal or spatial analyses in these cases is impossible without additional information.

looking closely at the data still holds. First, examine the data as a function of x. Are there specific regions where the values of the elements change abruptly? This could point to problems with the measurement instrumentation or places where the system has undergone a drastic change, making it difficult to compare elements. Check again for outliers, especially whether they follow any patterns in x. In many cases it is also useful to examine the data plotted by element index, i. Patterns that emerge in i can often point to problems with instrumentation, such as a bug that occurs every fifth measurement.

5.2   Element Spacing

Usually, the measurement locations x_i correspond to a spatial or temporal dimension. In many cases, the intervals between the measurement locations x_i dictate the analytic techniques that can be appropriately applied. In this section, I will describe a few categories of spacing so that you can develop some intuition for their differences. Figure 5.1 illustrates the first three of these categories visually.

5.2.1   Evenly Spaced Elements

Evenly spaced elements are the easiest type of 1-dimensional data set to analyze. The measurement locations in an evenly spaced data set follow the simple rule:

$$x_i = x_0 + ih, \tag{5.1}$$

where h is the spacing between the locations. Using this definition, the last location in the data set is at $x_{n-1} = x_0 + (n-1)h$. Evenly spaced data sets have two characteristic scales: the spacing between the points and the total length of the data set, $(n-1)h$. As I will demonstrate in Section 5.8, these characteristic scales influence the frequencies of the signals you will be able to detect.

Statistics like the mean, median, standard deviation, and mean absolute deviation can be reliably applied in the case of evenly spaced 1-dimensional data sets. Of course, the same precautions relevant to using these techniques on lists also apply to using them on 1-dimensional data.

5.2.2   Evenly Spaced Elements with Gaps

Imagine a thermometer that records the temperature once per second over the course of a day, except for an occasional few minutes when the power fails and the thermometer records nothing. The analyst has no knowledge of what the

Evenly Spaced Elements

Evenly Spaced Elements with Gaps

Unevenly Spaced Elements

FIGURE 5.1
Symbolic representations of three types of element spacings. The horizontal lines represent the dimension of the measurement locations, and the vertical hatches represent the locations in that dimension where elements are measured.

temperature was during these outages; the data are essentially missing.[2] The data set produced by this instrument will have evenly spaced measurements, separated by $h = 1$s, with gaps corresponding to the power outages. This is a 1-dimensional data set that has evenly spaced elements with gaps. A characteristic scale in this type of data sets is the length of the time the instrument is recording measurements without interruption, t. Analyzing trends in the data on shorter timescales than t can proceed as for evenly spaced elements, but if the behavior of interest involves a timescale longer than the instrument is continually online, some special techniques to deal with the missing elements are necessary. I'll discuss this further in the examples below.

Another example of a data set that has evenly spaced elements with gaps is the price of a company's stock on a stock market. Imagine the following system (this is a very simplified interpretation of how the actual market works): while the market is open, the price of the stock is updated regularly, but while the market is closed, there is no information on how much the stock is worth. In this picture of stock prices, after the market closes, news about the company or trends in foreign markets may alter the value of the stock, but there won't

---

[2]I could estimate the temperature during the outage assuming it varied smoothly, but these estimates cannot replace actual measurements.

be another measurement of the stock's value until the market opens in the morning.[3]

Calculating simple statistics for data sets that have evenly spaced elements with gaps requires judgement, because of the potential for bias. A key piece of information is whether the data outages are random. If they are, you should be able to calculate the mean, median, etc., without being contaminated by bias. However, if the data are missing due to the value of the behavior of interest, that's a clue that your statistics could potentially suffer from systematic error. For example, imagine a thermometer that shut down and didn't record data whenever the temperature exceeded 100°F. Calculating the mean of the recorded data will result in a statistic that is biased too low. Outages directly related to the value of the data aren't the only situation where you can get biased statistics: imagine an experiment where the temperature was measured every minute for the length of a day, except for the period from 1-2 p.m. when the observer went to lunch. In this situation, the reason for the missing data is independent of the value of the measurement, but it so happens some of the highest temperatures during the day are not recorded. An average calculated from these data would still be biased. As always, think before you calculate!

5.2.3   Unevenly Spaced Elements

Data sets with <u>unevenly spaced elements</u> (those without a simple regular pattern) are the most difficult to analyze. One tactic is to use a technique like interpolation (see Section 5.4) to map the elements onto an evenly spaced grid and proceed with analysis for evenly spaced elements. This would be less than ideal because it requires altering the original form of the measurements, but it is occasionally necessary for making progress. Later, I will also discuss techniques that treat the original unevenly spaced elements directly.

The same guidance I gave for detecting bias in data sets that have evenly spaced elements with gaps applies to unevenly spaced elements. What is causing the uneven spacing? Are there important locations in the data set where the sampling is sparse? Again, use your best judgment, and when you present your analysis, explain potential shortcomings.

5.2.4   Circular Data Sets

An interesting situation that can occur with a data set that has any of the above three element spacings is for the last element in the sample to loop around to a location near the first element in the structure; these structures are <u>circular data sets</u>. They are fundamentally different from linear data structures and require special attention. The archetypical setup for circular data sets is to label the measurement location by degree number around the circle (up to 360° obviously), or equivalently to use radians (up to 2π). Circular data

---

[3]In practice, stock futures address this issue, but let's not worry about that here.

sets often do not have an unarbitrary location to place the origin of the coordinate system; in other words, which element should be labeled $x_i = 0°$? In this situation, the analytical techniques utilized should give the same results when the origin is shifted.

5.3   Three Example 1-Dimensional Data Sets

In this chapter, I will examine three 1-dimensional data sets, each with a different structure. These data sets can be downloaded from the book's website, www.applyinganalytics.com.

5.3.1   Voltage Measured by a Sensor

The first example 1-dimensional data set is the voltage measured by a sensor at regular time intervals. The changing voltage is a result of an unknown mechanical process that an analyst is trying to understand better, and the measured voltage is the window into its internal workings. Data this well structured is hard to come by; unsurprisingly, it is fake. The data set consists of 1000 points measured from $x_0 = 0$s to $x_{999} = 999$s with no error bars. Figure 5.2 plots the voltage as a function of time. I'll call this the voltage data set.

What can be learned from simply examining the data by eye? The voltage appears to complete 10 oscillations over the length of the data set, with an amplitude around 4 or 5V. Since the element values don't line up perfectly, there appears to be some noise in the graph. Judging from the scatter in the points around the peaks, the noise amplitude is approximately 1V. The minima and maxima of the peaks and valleys are approximately symmetric, but the amplitude appears to vary slightly throughout the data set. By eye, it's hard to say what could be causing that effect.

Calculating the elementary statistics of the voltage data set confirms the initial impressions. The mean of the data is -0.05V, which, compared to the amplitude of the signal, is a relatively meaningless number in this situation. If the length of the data set had been extended another half oscillation, the mean would be somewhat larger than zero, without any real change in the structure of the data. The same is true for the median. The standard deviation is approximately 2.5V, which is expected given the estimated oscillation amplitude. (The standard deviation of elements measured at regular intervals from a simple sine wave is the amplitude of the sine wave divided by $\sqrt{2}$.)

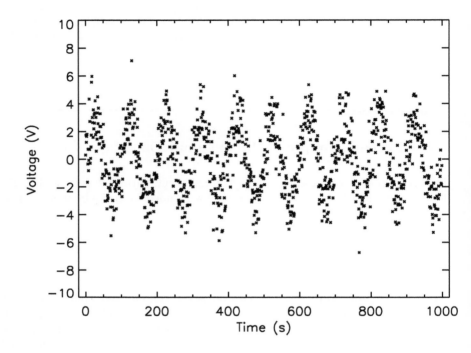

FIGURE 5.2
Voltage measured by a sensor as a function of time.

5.3.2   Elevation Data

The second example 1-dimensional data set I will examine is the elevation
as a function of longitude all the way around the Earth's equator. This is
an evenly spaced circular data set, with about 2×10^4 elements. The data
themselves come from the 1 arcminute[4] resolution Earth Topography Digital
Dataset (ETOPO1), published by the U.S. National Geophysical Data Center
(Amante and Eakins 2008). The ETOPO1 is a composite map of the entirety
of the earth's surface constructed from a variety of sources, including satellite
radar imagery and soundings from nautical vessels. The measurement error in
each of the elements is essentially zero. From this 2-dimensional data set, I have
extracted the loop around the equator. At the equator, a degree corresponds
to about 11 km, so the elements have a spacing of about 1.9 km. I'll call this
the elevation data set.

Figure 5.3 plots the elevation at the equator as a function of longitude.
Negative longitude values correspond to the Western hemisphere, and positive

---

[4]An arcminute is $1/60^{\text{th}}$ of a degree.

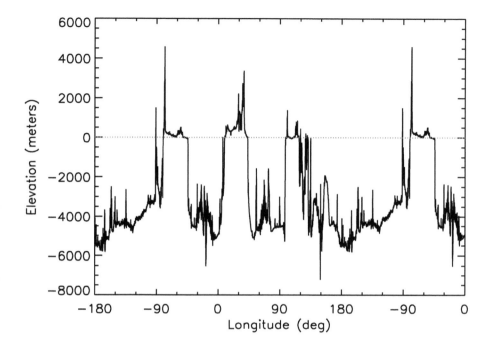

FIGURE 5.3
Elevation as a function of longitude along the equator. Negative longitudes are in the Western hemisphere, and positive in the Eastern. Note that the plot spans 540° of longitude, with the Western hemisphere data repeated to make it easier to trace features around the circle.

to the Eastern. The three large landmasses on the equator are South America (−80° to −45°), Africa (10° to 45°), and Indonesia (90° to 135°). The highest point on the equator is at Volcán Cayambe, a volcano located near −78° in Ecuador. The deep trench in the Atlantic Ocean is the actual boundary between tectonic plates at the Mid-Atlantic ridge. The trench in the Pacific is a similar feature located near Papua New Guinea.

Approximately 21% of the data points are located above sea level, with a mean of −2901 m and $\sigma_{n-1} \approx 2100$ m. These simple statistics should be interpreted carefully, because, to my eye, the frequency distribution of the elements looks bimodal; there are many points located near 0 m elevation, and many located at around 4000 m below sea level. For the reasons I discussed in Section 3.4.4, the standard deviation isn't going to be very informative in this situation. To confirm the bimodal nature of the data, you can plot a histogram of the number of points binned by elevation. The larger number of points below sea level is also evident in the value of the median, −3835 m.

5.3.3   Temperature Data

The third example 1-dimensional data set I will examine is the average daily temperature in Addis Ababa, the capital of Ethiopia.
This data set was collected by the National Climactic Data Center
(www.ncdc.noaa.gov/oa/ncdc.html), which I accessed through the University
of Dayton archive (www.engr.udayton.edu/weather). The average temperature for a single day is the mean of 24 daily readings. Occasionally, malfunc-
tions in the instrumentation caused readings to be missed; days on which this
occurred are flagged as elements with a value of '-99'. This is an evenly spaced
data set with gaps that I'll refer to as the temperature data set.

The data set spans from 1 January 1995 to 26 March 2009, a total of
5199 elements (Figure 5.4). Of these, there are no data for 1688 days. There
are gaps throughout the data set; there is a large, sparsely populated gap
approximately 2 yrs in length, from July 2002 to November 2004. For the
data that are present, $\mu = 63°F$, the median is $62.6°F$, $\sigma_{n-1} = 3.5°F$, the
maximum is $77.0°F$, and the minimum is $51.9°F$. Keep in mind that these
are daily average temperatures, so the maximum and minimum values do not
correspond to the full range of recorded temperatures in Addis Ababa.

Admittedly, this data set breaks one of the cardinal rules; it is a step removed from the measurements that are recorded from the instruments themselves due to the 24 hr averaging. This data processing step will hide many
trends (such as, obviously, the day/night temperature cycle) but this is the
only version of the data set I have, so let's see what can be done with it.

5.4   Interpolation

Frequently, an analyst needs to estimate ψ at a value of x that falls in between two measurement locations x_i. This process of approximation is called
interpolation. Let's be clear; this is an estimate of ψ, since ψ could take a
wide range of values depending on the variability of the behavior of interest
over that spatial range. In other words, who knows what ψ is doing when
you aren't looking? By making some assumptions about the shape of ψ, you
can come up with an estimate. Interpolation can be used on data sets with
any element spacing, though the structure of the data set can impact which
particular interpolation method is best.

Interpolation can also be used to manipulate the structure of data sets.
For example, you could take an unevenly spaced data set and interpolate it
onto an evenly spaced grid. However, procedures like this should only be used
as a measure of last resort. If the questions you are interested in answering
do not require a restructuring of the data, just use the original data (I cannot

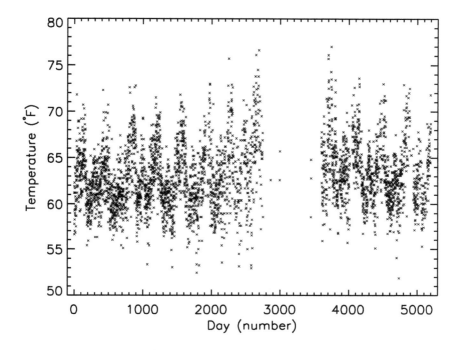

FIGURE 5.4
Average daily temperature in Addis Ababa, Ethiopia, from 1 January 1995 to 26 March 2009.

emphasize this enough). You will save yourself a lot of headaches in the long run by staying as close to the originally measured data as possible.

How does interpolation work? Let's say I want to estimate the value $\widetilde{\psi}$ at $\widetilde{x}$ in between x_i and x_{i+1}, which have corresponding values ψ_i and ψ_{i+1}. There are a lot of different algorithms for actually performing an interpolation, and they give different estimates. The shape of the signal in the data set determines which of the algorithms is best. In other words, a straight line connecting the two adjacent points won't always give the best estimate.

Here, I'll discuss an algorithm called <u>linear interpolation</u>, which is based on exactly that method: drawing a line in between ψ_i and ψ_{i+1} and finding where $\widetilde{\psi}$ lies on that line. This is about the simplest estimation technique imaginable. In pseudocode, the algorithm for this linear interpolation is:

slope$= (\psi_{i+1} - \psi_i)/(x_{i+1} - x_i)$
$\widetilde{\psi} = \psi_i + \textbf{slope} \times (\widetilde{x} - x_i)$

Figure 5.5 depicts this procedure in schematic form. This method, since it

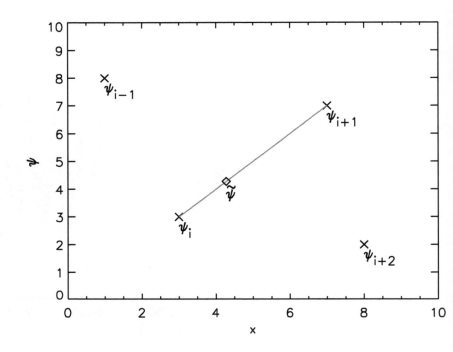

FIGURE 5.5
A schematic representation of linear interpolation for a data sets with unevenly spaced elements. The x's represent elements in ψ, the grey line is the
interpolation between ψ_i and ψ_{i+1}, and the diamond is the value at the point
$\tilde{x}$ where I need the estimate, $\tilde{\psi}$.

only relies on the adjacent values of x, can be used for 1-dimensional data sets
with any element spacing. Of course, this type of interpolation will result in
sharp corners and discontinuous derivatives at the measurement locations x_i,
which, for most functions, is not representative of reality.

It is easy to imagine other algorithms to perform interpolation. Two
straightforward ones involve fitting quadratic curves to the nearby points. A
simple quadratic interpolation is to solve for the second degree polynomial that
connects ψ_{i-1}, ψ_i, and ψ_{i+1}. Alternatively, you could do a linear least squares
fit of a second degree polynomial to the four points $\psi_{i-1}, \psi_i, \psi_{i+1}$, and ψ_{i+2}.
For both of these algorithms, $\tilde{\psi}$ is placed in between ψ_i and ψ_{i+1}. Note that
although these algorithms are (slightly) more sophisticated than linear interpolation, they don't solve the issue of the interpolated derivatives being
continuous at the points ψ_i, since the quadratic changes discontinuously for

$\widetilde{\psi}$ on opposite sides of a measurement location. However, at least these algorithms stop the interpolated values themselves from having sharp corners.

There are certainly occasions when you'll want the interpolated quantities to have both smooth corners and continuous derivatives. In these cases, your best bet is a technique called spline interpolation. I won't discuss the algorithm for splines, but instead refer you to Kreyszig (1993) and Press et al. (2007).

Like all of the quantitative techniques we discuss, interpolation should not be used everywhere. As an example of a place where interpolation would not be appropriate, consider a large gap in the temperature data set. The largest single gap spans more than a year, between 17 March 2003 and 13 June 2004. Would a linear interpolation between these two points be appropriate? Of course not. This data set clearly has an oscillatory component with a period smaller than the gap in the data, and a linear interpolation would grossly distort the likely trends in the data. On the other hand, gaps in the temperature data that are a few days or shorter are appropriate candidates for interpolation, because the large-scale trend only completes a small fraction of its oscillation over that range. So, in general, when the gap between x_i and x_{i+1} is much smaller than the periods of the major oscillations in the data, then interpolation should give a fair estimate of ψ.

5.5 Smoothing

Noise and variability in a data set can hide the signal an analyst is interested in identifying. Fortunately, there are a variety of algorithms available to make the signal stand out. In this section, I will discuss several that fall into the category of smoothing, a technical term used for techniques that minimize the influence of noise and variability.

5.5.1 Running Mean

The running mean is a smoothing technique useful for revealing trends in noisy data sets. The running mean is particularly effective when dealing with noise that averages to zero, like noise that is drawn randomly from a Gaussian distribution. Conceptually, the value of the running mean, $\mathring{\psi}_i$, is equal to the mean of the elements nearby to ψ_i. In pseudocode, for evenly spaced data, the algorithm for the running mean looks like:[5]

---

[5]For readers who don't know how to program, a **for** loop repeats the statement inside the brackets once for each value of the index it is run over.

for $i = 0$ **over** $i \leq n - 1$ {
$$\mathring{\psi}_i = \textbf{mean}(\psi_{i-w_h} : \psi_{i+w_h})$$
}

where w_h is the half width, in index form, of the region the running mean is averaged over. The loop runs from the start of the data set ($i = 0$) to the end ($i = n - 1$). The size of the region that the average is calculated over, known as the window, is a subjective decision; in fact, using different size regions will yield different running means and will emphasize different aspects of the data. A wider window tends to be more stable and less sensitive to outliers than a narrower window because its mean is calculated using a larger number of elements. I often will examine running means with varying window sizes for a single data set to see the different features that emerge.

Readers paying close attention will notice a problem with the above algorithm; elements near the edges of the data set will have imbalanced running means, because there are no points at $i < 0$ and $i > n-1$ so the window will be asymmetric. Elements that meet this criteria fall in the zone of influence of the edge. The safest course of action is to discard these elements from the running mean, such that $\mathring{\psi}_i$ is only valid for $w_h \leq i \leq n - 1 - w_h$. You could calculate the running mean for elements in the zone of influence using a smaller w_h, but I would not recommend that approach because mixing elements with different running mean windows in the same ψ_i makes for a confusing interpretation. The same logic also applies to evenly spaced elements with gaps; elements which fall in the zone of influence of a gap should also be discarded. If the gaps in the data are closely packed, a running mean with a large window may be a useless statistic, since most elements in the data set will be in the zone of influence of one gap or another. In that case, consider using the techniques for unevenly spaced elements, described below.

The running mean can suffer from all of the potential difficulties associated with the standard mean described in Section 3.3.1. If there are elements that are outliers, they can dominate the running means of the surrounding points. If the frequency distribution of elements is bimodal, the running mean can result in intermediate and misleading values. Keep an eye on which elements are dominating the running mean in each region of your data set.

The running mean is straightforward to modify for circular data sets; for points within w_h of $i = 0$ or $i = n - 1$, include the ψ_i from the other end of the indexing scheme in the running mean. For readers that are coding their own algorithms, this can be done by adding an **if-then** statement into your code. One nice feature of evenly-spaced circular data sets is that there are no edge effects.

The running mean can be applied to all three of the example 1-dimensional data sets. Let's use it on the simplest case, the voltage data set. Figure 5.6 shows the results when a running mean with $w_h = 5$ is applied to the data. This technique is particularly effective on this data set because the noise has a mean of 0. (I know this because I added the noise to the fake data set.) Notice

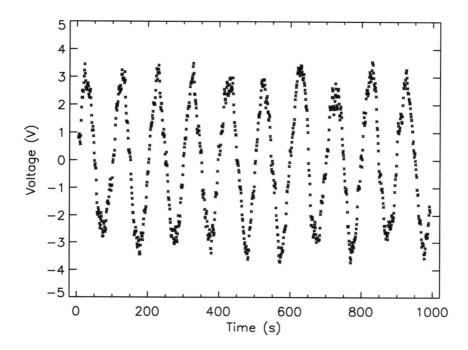

FIGURE 5.6
Running mean with $w_h = 5$ applied to the voltage data set.

how minimizing the effect of noise reveals that the peaks of the oscillation are uneven, implying that there is more going on here than a simple single frequency oscillation. Further investigation later in this chapter will clarify this structure.

For unevenly spaced data, calculating the running mean is not so simple, because if you shift w_h points to the left, you are unlikely to cover the same distance in x as traveling w_h points to the right. This means that the window over which the running mean would be calculated is no longer symmetric, and also will cover a different range in x as a function of i. These effects would lead to a statistic that is difficult to interpret. The formally correct technique to apply to this situation is Savitzky-Golay filtering (Savitzky and Golay 1964; Press et al. 2007). However, the simplest analog to an index-based window is one with dimensions prescribed in x rather than in i. This window will likely result in a unbalanced running mean, with unequal number of points on the left and right side of the window, which is a poor practice in general. Therefore, be extremely careful using this technique; use it only when you need a fast version of a running mean with minimal coding time, because it's likely to be biased and inaccurate.

5.5.2   Median Smoothing

Readers of Chapter 3 know what comes after a technique that relies on the mean: one that relies on the median! Median smoothing, conceptually and algorithmically, operates similarly to the running mean. It is a technique appropriate for data sets with large noise and variability, particularly if those influences don't average to zero.[6] The pseudocode algorithm for median smoothing can be written as:

for $i = 0$ **over** $i \leq n - 1$ {
$\quad \check{\psi}_i = \textbf{median}(\psi_{i-w_h} : \psi_{i+w_h})$
}

where w_h is again the half width, in index form, of the median window. The same warnings regarding edge effects in the zone of influence for the running mean apply to median smoothing.

Using median smoothing on the elevation data set removes the impact of the high mountains and deep ridges (Figure 5.7). For this analysis, I've chosen $w_h = 150$. Note how median smoothing highlights the general shape of the continents, but obscures some important features. For example, the ridge just to the west of South America near $-90°$ longitude does not rise above sea level after the median smoothing has been applied. This happens because in the original data set there are only a few elements above sea level. Running another median smoothing with a smaller w_h is recommended if you are interested in features of this scale. As a general rule of thumb, features smaller than $2hw_h$ will not appear in median smoothed data. The same rule does not apply to the running mean, since even one extreme valued point can strongly influence that smoothing technique.

Unevenly spaced data sets are not amenable to analysis with median smoothing. If you must use median smoothing on one of these data sets, you may be forced to interpolate the unevenly spaced data onto a regular grid.

5.6   Fitting 1-Dimensional Data Sets

It is often helpful to find an equation to approximate a 1-dimensional data set. The process of determining an approximation is called fitting or, equivalently, regression. You can think about fitting as akin to drawing a smooth line or curve to approximate ψ. Fitting a 1-dimensional data set is part of a larger set of fitting problems I will discuss in this book; it's also connected to some of the ideas from hypothesis testing I discussed in Section 3.6.

Why is fitting an equation to a 1-dimensional data set useful? Well, data

---

[6]Noise with extreme high and low points is sometimes known as salt and pepper noise.

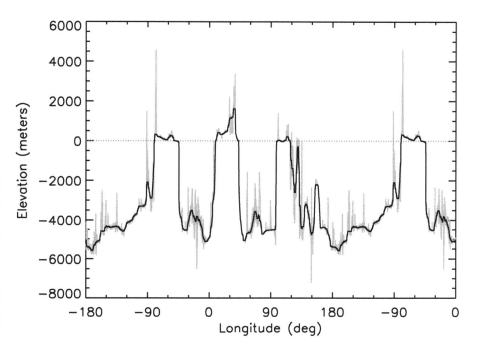

FIGURE 5.7
Elevation data for the earth's equator after median smoothing applied with
$w_h = 150$. The original data is plotted with a solid grey line. The horizontal
dotted line is sea level.

sets made from real data tend to have noise, so it can be helpful to have a
smooth approximation of ψ to make predictions about future measurements
or for use in modeling. There are also occasions when the 1-dimensional data
set is computationally difficult to work with. Using the temperature data set
as an example, an analyst might need to approximate the difference between
the average daily temperatures on June 1, 2005 and December 1, 2005. If you
had an approximate fit for the temperature data set, this difference could be
estimated easily, without having to go back to the original data.

Fitting is quite different from the analytical techniques I've discussed to
this point; there is a lot more flexibility and, therefore, analytical judgment
and intuition required. When working with the mean and the median, for
example, the algorithms were straightforward but judgment was required to
interpret the outputs. In fitting, judgment is required much earlier in the
process, when deciding on the type of equation to fit as well as in the interpretation of the fit's outputs.

In fitting, the term dependent variable refers to the variable that is being fit, and the term <u>independent variable</u> refers to the variable that the dependent variable is a function of. In a 1-dimensional data set, ψ_i is the dependent variable and the measurement locations x_i are the independent variable. I'll call the values output by the fit $\bar{\psi}_i$, to make clear that they are of the same dimension as ψ_i and are measured at the same locations.

So, how is fitting actually done? In practice, when performing a fit, an analyst first chooses a <u>functional form</u> that gives the fit a broad shape, like a parabola, a line, a sine wave, or a bell-curve. The functional form contains a set of <u>parameters</u> that control the more precise shape of the functional form (see the inset on linear and nonlinear parameters). Next, some algorithm is used to adjust the parameters, stretching or distorting the fit's shape so that it better matches the data. For example, is the bell curve narrow or wide? Is the slope of the line steep or flat? The parameter values settled on should be ones that result in a fit that is a good match to the data.

As I mentioned earlier, choosing the functional form for a fit requires an analyst to exercise judgment. You can fit a data set perfectly with a functional form that has as many parameters as there are elements, but then each of the parameters wouldn't mean very much. Simpler functional forms are generally preferred, but not so simple that important features in the data are omitted. I'll admit this is a tough balance to strike; there isn't a right answer and choosing a good one is difficult even for experienced analysts. Still, choosing an appropriate functional form is a skill you will learn from practice.

To determine if one particular combination of parameters produces a better fit than some other combination of parameters, I need a metric to quantify the quality of a fit. After all, I'm trying to determine which combination of parameters produces the best fit. In the hypothesis testing discussion (Section 3.6), the metric χ^2 served this purpose. A simple analogous metric that can be used in fitting is S^2, given by:

$$S^2 = \sum_{i=0}^{n-1} \left(\psi_i - \bar{\psi}_i\right)^2 , \qquad (5.2)$$

Like the χ^2 metric, S^2 sums the squares of the residuals between the data and the predicted value.[7] Fits that depend on minimizing S^2 are known as <u>least-squares</u> fits.

---

[7]There is also a χ^2 statistic for 1-dimensional fits which incorporates the uncertainty in the measured elements. I won't discuss it in this book, but Taylor (1997) has more detail.

Linear vs. Nonlinear Parameters

Parameters in a fit's functional form can be either linear or nonlinear. The distinction is extremely important, because fitting functional forms that contain only linear parameters is a good deal more straightforward than fitting functional forms that contain even one nonlinear parameter. However, nonlinear parameterizations can provide additional flexibility. Basically, you can fit whatever function you want, but the choice you make has a strong influence on the methods appropriate to find the best fit. Some software programs may not be able to handle some fitting methods, so the simpler you can keep the fit, the better.

So, how can you distinguish between linear and nonlinear parameterizations? In general, linear parameters are found in front of a function, whereas nonlinear ones are nested inside. Some examples of linear parameterizations are:

$$\bar{\psi} = Ax, \quad \bar{\psi} = Bx^2, \quad \bar{\psi} = Ce^x + Dx, \quad \bar{\psi} = E^2 \sin(x). \tag{5.3}$$

In these equations, the fit parameters are capital roman letters. The dependent variable is $\bar{\psi}$ and the independent variable is x. Notice how changing the value of the parameters changes the shape of the function. For example, changing A controls the slope of the fitted line. Take a look at the last function in this list – why is E^2 a linear parameterization? Well, by making the substitution $F = E^2$ the equation could be changed to $\bar{\psi} = F \sin(x)$, which is a simple linear parameterization.

Examples of nonlinear parameterizations include:

$$\bar{\psi} = \sin(Gx), \quad \bar{\psi} = e^{Hx}, \quad \bar{\psi} = I \sin(Jx). \tag{5.4}$$

The final equation in this list is linear in I but nonlinear in J.

Fitted functions often additively combine several separate functional forms, like the third example in Equation 5.3. When functional forms are added, if any of the fitted parameters is nonlinear, the entire fit is nonlinear.

Sometimes, by being clever, you can turn a nonlinear parameterization into a linear one. This is important because, as I've emphasized, linear fitting methodologies are a good deal simpler than nonlinear ones. For example, by taking the logarithm of both sides of the second equation in Equation 5.4 and substituting $z = \ln \bar{\psi}$, I find $z = Hx$. This is a linear parameterization which can be fit after taking the logarithm of the elements ψ_i.

When calculated for the best-fit combination of parameters, the value of S^2 is connected to how well the parameters have been determined. This is known as the uncertainty in the fitted parameters, a topic I will discuss in more detail in the following sections. For now I will note that the connection should not come as a surprise. The larger the residuals, the more the parameters could be adjusted and still get a similar value of S^2.

The computationally difficult part of fitting is usually determining what combination of parameters for a given functional form will minimize S^2. For functional forms with fully linear parameterizations, the equations for the best fit parameters can be solved in general form (see Section 5.6.2). For non-linear parameterizations, an algorithm generally must search around trying various combinations of parameters. Fitting nonlinear functions is, in general, computationally intensive and fraught with complexity. That's why I'm only discussing the topic here generally. An excellent introductory discussion of how to carry out these fits can be found in Bevington and Robinson (2003), and more advanced material in Press et al. (2007).

Minimizing S^2 isn't a technique that can be used blindly; if the functional form you choose doesn't match the shape of the data, minimizing S^2 doesn't mean much. For example, if you have a frequency distribution that is parabolic in shape, you can fit a sine curve to it, but you're not going to learn much from that fit. Keep in mind that the algorithm will happily tell you how well you've determined the functional form's parameters, and those numbers could be perfectly reasonable, even if the functional form you have chosen is a completely inappropriate shape! So your analytical judgment is important when choosing the functional form. Furthermore, like the variance, which has a similar form, S^2 has the potential to be dominated by outliers, meaning that a combination of parameters can produce a fit that matches most of the frequency distribution very well, but have a large S^2 because of a few outliers. However, in most circumstances, fits with lower S^2 have smaller differences between the fit and the frequency distribution.

Now that I've covered the basics of fitting 1-dimensional data sets, let's proceed by discussing some examples. In the next subsection, I'll cover the unusually simple example of fitting a straight line. Then, as an advanced topic, I'll cover the general case of fitting any linear function via least squares.

5.6.1   Straight-Line Fitting

One of the simplest functional forms to fit is a straight line. A straight line has two linear parameters: the slope and the y-intercept. In the general notation we are using in this section, the equation for a straight line can be written as

$$\bar{\psi} = mx + b, \tag{5.5}$$

where m is the slope and b is the y-intercept. Finding the best fit straight line means finding the values of m and b that best fit the data.

A straight-line fit is a special case of linear fitting in that the best-fit solution that minimizes S^2 reduces to a simple set of equations that are calculated from the data. The slope of the best-fit line is given by:

$$m = \frac{n\left(\sum_i x_i \psi_i\right) - \left(\sum_i x_i\right)\left(\sum_i \psi_i\right)}{n \sum_i x_i^2 - \left(\sum_i x_i\right)^2}. \tag{5.6}$$

Once you've calculated the best-fit slope, you can use it to calculate the best-fit y-intercept:

$$b = \mu_\psi - m\mu_x, \tag{5.7}$$

where μ_ψ is the mean of ψ and μ_x is the mean of x.

The equation for the best fit slope is not easy to make sense of; there are quite a few variables there. The numerator is the difference between two terms: the first term sums the product of the independent and dependent variables and the second term multiplies the sums of the two variables. To better understand how these two terms interact, consider what would happen if you were fitting a data set where ψ was constant. In this limiting case, the two numerator terms would exactly cancel, leading to a best-fit slope of zero, as you would expect. In the general case, you can think about the first term as capturing the relationship between the two variables, and the second term as correcting for the contribution from the product that would result even if ψ was constant.

The equation for the best-fit y-intercept is easier to understand. It is simply the difference between the mean of the dependent variable and the mean of the independent variable times the best-fit slope. It's essentially the constant factor that is left after the slope has taken care of its part of the relationship between the two variables.

How well have the slope and y-intercept been determined? In other words, how precisely have the data constrained the values of the fit parameters? In the special case of the straight-line fit, this question can also be answered with a simple algebraic form for the uncertainty in the parameters. These uncertainties are given as 1-σ error bars. The uncertainty in the best-fit slope, written as σ_m, is:

$$\sigma_m = \frac{S}{\sqrt{n-2}} \sqrt{\frac{n}{n \sum_i x_i^2 - \left(\sum_i x_i\right)^2}}, \tag{5.8}$$

where S is the square root of the sum of the squared residuals, S^2, described in the previous section. The dependence of σ_m on S^2 illustrates that when the data values are near to the fit values, the parameters are well-determined. When the fit is not of high quality and S^2 is large, the parameters are relatively poorly determined. The uncertainty in the y-intercept is given by:

$$\sigma_b = \frac{S}{\sqrt{n-2}} \sqrt{\frac{\sum_i x_i^2}{n \sum_i x_i^2 - \left(\sum_i x_i\right)^2}}. \tag{5.9}$$

A convenient place to perform our first fit is the age list, introduced in Chapter 3, at the falloff in population that begins near 45 yrs old. Note that even though I am fitting the frequency distribution of the age list, this is different than hypothesis testing a distribution. Instead, I'm fitting the absolute number of people by age, which is a 1-dimensional data set in its own right. The x's in Figure 5.8 are the raw data of the age list without the coarse

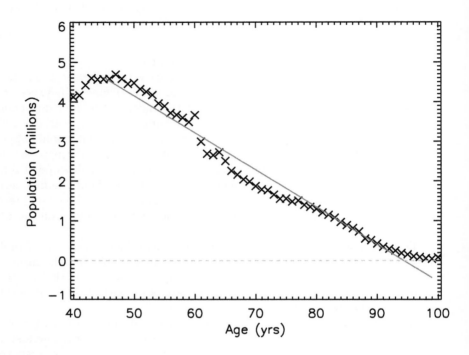

FIGURE 5.8
The U.S. population as a function of age in November 2007, from 40 to 100 yrs (x's). The fit to the portion of the population ages 45 to 99 yrs is overplotted as a solid grey line. The zero population level is shown as a dashed light grey line. Notice how the fit produces nonsensical negative populations at large ages.

grained ten-year bins of Figure 3.2. This falloff is generally smooth, but there are a few noticeable jumps. The jumps around 60 yrs and 87 yrs likely are due to the two World Wars America fought, and the baby booms that followed the return home of the soldiers. Notice also the small uptick in the point at 100 yrs, which is an artifact of binning everyone older than that threshold. I'll restrict the fit to between the ages of 45 to 99 yrs, starting where the falloff begins and ending just before the artificial uptick at 100 yrs.

How do I place the data into a 1-dimensional data structure? First, I will specify the independent variables: $x_i = 45$ yrs, 46 yrs... and the dependent variables $\psi_i = 4.6 \times 10^6$ people, 4.6×10^6 people... (the number of people in the United States of those ages). Including the endpoints of the fit range, $n = 55$. Plugging those values into Equations 5.6 and 5.7, I find the best fit values are $m = -9.4 \times 10^4$ people/yr and $b = 8.8 \times 10^6$ people.

Figure 5.8 shows the age list fit plotted on top of the frequency distribution. Some features of the fit stand out. First, the fit tracks the shape of the data fairly well, and the major features of the falloff are captured. Second, the high end of the fit falls so much that very high ages have a negative population, which makes no sense. In fact, this is a result of choosing a straight-line functional form that doesn't match the shape of the data, particularly the leveling off in the curve at high ages. This could have been avoided by choosing a more appropriate functional form or by fitting the data over a more limited range. Without looking at the details of the fit, there would be no way to know that the procedure produced a nonsensical answer for parts of the fit's range, so it is always necessary to check the fit against the original data.

The uncertainties in the fitted parameters are given by $\sigma_{p,m} = 2.1 \times 10^3$ people/yr. and $\sigma_{p,b} = 1.6 \times 10^5$ people. It's valuable to compare the magnitude of the errors to the parameters themselves; in this case, $|\sigma_{p,m}/m| = 0.02$ and $|\sigma_{p,b}/b| = 0.02$. From looking at the fit plotted on top of the data, the magnitude of the uncertainty makes sense, because the slope and y-intercept of the line are pretty well constrained by the data. More details about the specifics of straight-line fits, as well as additional examples, can be found in Taylor (1997).

I've seen many analysts apply a straight-line fit to data that isn't well fit by that shape because it's the only fitting method they know. Be careful not to fall into this trap. It's better to admit that performing an accurate fit is beyond your expertise than to fit an incorrectly shaped function and use the results from that fit. Inaccurate fit results don't do anyone any good.

5.6.2 General Linear Fitting (Advanced Topic)

The most reasonable computational method for linear parameter fitting via least-squares minimization is to solve a system of simultaneous equations using matrices. This section relies on a basic familiarity with linear algebra – it's the most advanced mathematics used in this book. I'm also going to assume that you can perform matrix arithmetic on a computer. Some popular programs with this capability are Mathematica™, MATLAB™, and IDL™. In keeping with the style of this book, I won't derive any of the results here, but simply help you to use them. For the derivations and other excellent advice, consult Heiles (2008).

In general, I would like to fit n_p linear parameters p_k in the form:

$$\psi = p_0 f_0(x) + p_1 f_1(x) + \ldots + p_{n_p-1} f_{n_p-1}(x), \tag{5.10}$$

where the $f_k(x)$ are the functional forms I am trying to fit, linearly summed.

It turns out that the combination of parameter values that minimizes S^2 can be found by solving a system of simultaneous equations. How can I find this solution? First, I define the $n_p \times n$ matrix of equation variables $\mathbf{F}$:

$$\mathbf{F} = \begin{bmatrix} f_0(x_0) & f_1(x_0) & \cdots & f_{n_p-1}(x_0) \\ f_0(x_1) & f_1(x_1) & \cdots & f_{n_p-1}(x_1) \\ \vdots & & & \\ f_0(x_{n-1}) & f_1(x_{n-1}) & \cdots & f_{n_p-1}(x_{n-1}) \end{bmatrix}, \tag{5.11}$$

and the n-vector of measurements $\mathbf{\Psi}$:

$$\mathbf{\Psi} = \begin{bmatrix} \psi_0 \\ \psi_1 \\ \vdots \\ \psi_{n-1} \end{bmatrix}. \tag{5.12}$$

The combination of parameters $\mathbf{P}$ that minimizes S^2 is given by:

$$\mathbf{P} = \left[\mathbf{F^T} \cdot \mathbf{F}\right]^{-1} \mathbf{F^T} \cdot \mathbf{\Psi}, \tag{5.13}$$

where "$\cdot$" denotes matrix multiplication, $^{\mathbf{T}}$ denotes the transpose, and $\mathbf{P}$ is the vector of linear parameters:

$$\mathbf{P} = \begin{bmatrix} p_0 \\ p_1 \\ \vdots \\ p_{n_p-1} \end{bmatrix}. \tag{5.14}$$

This vector will contain the parameter values for the best fit. I can calculate the fitted values, $\bar{\psi}_i$, by plugging the p_k into the fit equation and evaluating it n times. The difference between the measured and predicted values is then:

$$\delta\mathbf{\Psi} = \mathbf{\Psi} - \bar{\mathbf{\Psi}}, \tag{5.15}$$

and S^2 can be found using:

$$S^2 = \delta\mathbf{\Psi^T} \cdot \delta\mathbf{\Psi}. \tag{5.16}$$

This is just the vectorized form of Equation 5.2.

How precisely do I know the values of these parameters? The 1-σ error in the fitted parameters given by the vector σ_p can be found using:

$$\sigma_p^2 = \frac{S^2}{n - n_p} \, \mathbf{diagonal}\{[\mathbf{F^T} \cdot \mathbf{F}]^{-1}\}, \tag{5.17}$$

where the **diagonal** operator returns the elements along the diagonal of the matrix it operates on.

Often, analysts will use a modified version of S^2, where the original statistic is divided by a quantity called the number of degrees of freedom (a quantity

I briefly discussed in Section 3.6). This combination of variables appears at the front of Equation 5.17. The number of degrees of freedom is given by the number of elements in ψ, n, minus the number of parameters in the fit. This is the reason the quantity $\sqrt{n-2}$ appears in Equations 5.8 and 5.9.

This has been quite a bit of math; let's discuss how to take the information in the age list and set up a fit. I'll describe how to carry out the same fit to the age list I covered in the straight-line fitting section so you can easily see the differences between the two approaches. In this case, in terms of the general linear equation form (Equation 5.10), $p_0 = m$, $p_1 = b$, $f_0 = x$, and $f_1 = 1$. The matrices are given by:

$$\mathbf{F} = \begin{bmatrix} 45 & 1 \\ 46 & 1 \\ \vdots \\ 99 & 1 \end{bmatrix}, \tag{5.18}$$

and

$$\mathbf{\Psi} = \begin{bmatrix} 4.6 \times 10^6 \\ 4.6 \times 10^6 \\ \vdots \\ 4.7 \times 10^4 \end{bmatrix}. \tag{5.19}$$

Carrying out the matrix mathematics will give the exact same best fit and errors as we saw in the straight-line section; one of the exercises at the end of the chapter is confirm this result.

As a final example of 1-dimensional fitting, I will fit a single sine wave with a period of 100 s to the voltage data set. I fit the amplitude of this sine wave, but not the period because that would require a nonlinear fit. The fit is shown in Figure 5.9, plotted on top of the original data set. Notice how the sine wave does not appear to capture the behavior near the peaks very well – this is further evidence that there is some other behavior occurring in this data set that has not yet been revealed.

5.7 Increasing Local Contrast

The smoothing techniques described in Section 5.5 are used to reveal large-scale trends in the presence of small-scale variability. However, sometimes the analytically interesting features are in the smaller scale fluctuations rather than the larger trend. In these cases, smoothing may erase the features of interest, but other techniques are available that increase the local contrast and make the small-scale features more apparent.

One of these techniques is linked to the residuals, a concept I discussed

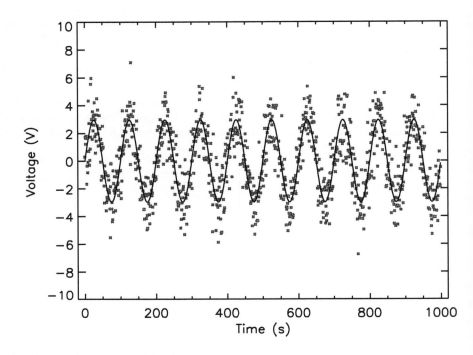

FIGURE 5.9
The voltage data set, plotted in greyscale x's, and a fit of a sine wave with a
100 s period plotted in a solid black line.

in the previous section on fitting. Imagine a 1-dimensional data set to which
you have fit a functional form, perhaps by minimizing S^2, the sum of the
squares of the residuals. Studying the residuals can be illuminating – simply
plotting the residuals often reveals subtle trends that were not noticeable in
the original data.

As an example of this technique in action, let's imagine you are trying
to determine whether a day in the temperature data set was hotter than
the typical temperature at that time of year. You could first fit an oscillatory
functional form with a period of 1 year to the data set to account for the annual
trend. The residuals left from that fit will then show the deviations from this
larger-scale trend – if the day you are interested in has a positive residual
(and the fit is a good match to the rest of the data), the day's temperature is
higher than the typical temperature for that time of year. When constructing
findings like this one, be aware of the variability and noise in the residuals and
don't overanalyze small fluctuations. One reasonable approach is to calculate
the standard deviation of the residuals and compare any elements of interest
to that value.

Other methods are available to increase the local contrast. For example, consider unsharp masking, a technique that originated in photography. The algorithm for unsharp masking proceeds by first creating a blurred or smoothed version of the data, then compares the original version of the data to the smoothed version. Usually, the smoothed version is subtracted from the original version. However, in data sets with a large dynamic range, it can be useful to calculate the ratio instead (e.g., Levine et al. (2006)). To use unsharp masking, you have to choose the size of the window used to create the smoothed image, as well as the smoothing technique.

5.8   Frequency Analysis

Frequency analysis is a form of pattern recognition used to identify repeating oscillations (or cycles) in data, commonly known as waves. Frequency analysis is also known as time series analysis in some fields.

Frequency analysis, of course, depends on frequency, a count of the number of oscillations a wave will complete in some amount of an independent variable, usually distance or time. The units of frequency depend on the units of the measurement locations array, x. For data that is taken as a function of time, the units of frequency are cycles/time. For data that is taken as a function of position, the units of frequency are cycles/distance. For circular data, frequency can be given as cycles/distance, cycles/degree, or cycles/radian, whichever is most appropriate.

I will use the symbol ν to represent the frequency of an oscillation, regardless of the actual units with which it is measured (f is also a common symbol for frequency). Oscillations that repeat relatively rapidly are high frequency, and oscillations that complete their cycles slowly are low frequency – the absolute values of which frequencies are high or low depend on the context of your analysis.

An example will help to illustrate how these ideas work. For one of the simplest wave functions, a cosine, frequency appears in the function multiplied by the independent variable and some constants, as in $\psi(x) = \cos(2\pi\nu x)$.[8] A cosine function completes a full cycle in 2π. Specifically, as x goes from 0 to $1/\nu$, the cosine completes an oscillation from its peak to its valley and back to its peak again. Not all repeating patterns are this simple, but sines and cosines serve as good examples and can be used to test your analytical routines.

Instead of using frequency, an alternative and equivalent method of characterizing cyclic patterns is in terms of the length (in time or in distance) a

---

[8]Careful readers may notice that the product of the frequency and the independent variable actually produces a number with the unit "cycles." Technically, the 2π inside the cosine function has the units "radians per cycle," because a cosine wave completes a cycle in 2π, and sinusoidal functions take inputs in radians.

full oscillation takes to complete; this quantity is known as the period, P, for oscillations in time or the wavelength, λ, for oscillations in space. The relation between these quantities and a wave's frequency is given by:

$$\nu = \frac{1}{\lambda} \text{ or } \nu = \frac{1}{P}, \tag{5.20}$$

with the first relationship applying to oscillations in distance and the second to oscillations in time.

Waves are also characterized by their amplitude, or strength, A. In the simple wave function, amplitude enters as a multiplicative factor in front of the wave's equation: $\psi(x) = A\cos(2\pi\nu x)$. The units of amplitude are essentially unrestricted.

For many wave functions, there is another variable in addition to the amplitude and the frequency: the phase. Think of the phase of characterizing the point in the oscillation that the wave is at when $x = 0$. The wave could be at a crest, a valley, or anywhere in between. Mathematically, the phase enters the form of the simple wave as the variable ϕ_0, with units of radians, in $\psi(x) = A\cos(2\pi\nu x + \phi_0)$. The phase will also appear with a minus sign in front of it – whether or not the minus sign is present is a convention, meaning it is an arbitrary choice and which one you choose should not affect the results of your analysis.

Figure 5.10 depicts a notional data set of a simple sine wave pattern. The signal apparent in this data set has a wavelength of 10 s, a frequency of 0.1 Hz,[9] an amplitude of 3 (no units given), and a phase of 0.

Instead of frequency, some analysts use a related quantity called the angular frequency, ω, which is equivalent to a standard frequency but is somewhat more difficult to make intuitive sense of. Conceptually, you can think of a wave as a point repeatedly moving around a circle. Movement around the circle corresponds to the oscillation traversing its peaks and valleys. In high angular frequency oscillations, the point moves around the circle relatively quickly, and in low angular frequency ones it moves relatively slowly. Completing half an oscillation, such as a sine wave moving from a maximum to a minimum, can be represented as traveling an angular distance π around the circle. This defines a straightforward relationship between the two types of frequency:

$$\omega = 2\pi\nu \text{ radians/cycle}. \tag{5.21}$$

As the units of frequency are cycles/time or cycles/distance, the units of angular frequency are radians/time or radians/distance. Pay close attention to the units of both types of frequency, because radians are rarely explicitly written out and it is very easy to confuse an angular frequency with a standard frequency. Be extremely deliberate when both quantities are being used in the same analysis, because if you make a mistake with the factor of 2π, it will change your results by almost an order of magnitude!

---

[9]The unit Hertz, abbreviated Hz, is shorthand for cycles/s. There is no analogous unit for cycles/m.

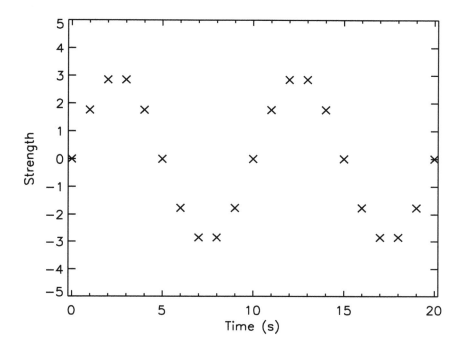

FIGURE 5.10
A simple wave pattern in a data set taken as a function of time, with $A = 3$.

I've discussed frequency analysis so far in terms of sines and cosines, but the repeating features in your data don't need to have these sinusoidal shapes to be amenable to frequency analysis. There are techniques that can be applied for oscillations of many different shapes, and your choice of technique should be influenced by the shapes of the oscillations in your data set. However, in this book, I will concentrate on sinusoidal oscillations because it so happens that sinusoidal features are common in nature and other observations. They are also mathematically straightforward.

When it comes to analyzing real data sets, there are some simple rules regarding the range of frequencies that can be reliably investigated. These rules apply regardless of the shape of the oscillations in your data because they only depend on the structure of x and not the element values in ψ. First, to make a defensible statement about an oscillation, you must observe it complete a full cycle. How else would you know it returns to the same value at which it started? Therefore, the period of the frequencies investigated must be smaller than the length of the data set, $|x_{n-1} - x_0|$. In the special case of circular data, this longest period is clearly $360°$. Second, a similar rule applies to the closest spacing between measurement locations. To reliably measure an oscillation, you need at least 3 points, corresponding to a peak, a valley, and a second peak. If the elements are spaced too far apart, the wave will complete its oscillation before the next measurement. Call the closest spacing in your measurement locations Δx; for evenly spaced data, Δx is also known as the sampling rate. In evenly spaced data, with or without gaps, $\Delta x = h$, but this is not the case for an unevenly spaced data set. You won't be able to reliably measure any oscillation that has a peak to valley distance less than Δx. Thus, the shortest wave period that can defensibly be measured, referred to as the Nyquist limit, is $2\Delta x$. Combining this limit with the first rule, the range of valid frequencies to investigate in a given data set is given by:

$$\frac{1 \text{ cycle}}{|x_{n-1} - x_0|} \leq \nu \leq \frac{1 \text{ cycle}}{2\Delta x}, \tag{5.22}$$

and for circular data:

$$\frac{1 \text{ cycle}}{360°} \leq \nu \leq \frac{1 \text{ cycle}}{2\Delta x}. \tag{5.23}$$

The circular data limits are given for ν with units of cycles/degree, but they could easily be written in whichever units you find most convenient. In the case of unevenly spaced data, you should apply some discretion when investigating frequencies that approach the Nyquist limit, since there may not be many measurements taken with that minimum spacing between measurement locations.

Looking again at Figure 5.10, what types of waves could be present in the system, but remain undetected? The sampling rate of the data, h, is 1 sample/s, and the data set is 20 s long. Therefore, periodic signals with a period shorter than 2 s will be difficult to investigate. The same is true for signals with a period longer than 20 s.

Unfortunately, frequencies outside the valid range of investigation can still have an impact on your data set. For example, Figure 5.11 demonstrates an unexpected mechanism for generating the signal observed in Figure 5.10. The overplotted function is given by $\psi(t) = 3\sin(-2\pi t \times 0.9 \text{ Hz}) = -3\sin(2\pi t \times 0.9 \text{ Hz})$. This wave has a period of 1.1 s, outside the Nyquist limit. I have no way of knowing whether the observed data were produced by this function or the more obvious $\psi(t) = 3\sin(2\pi t \times 0.1 \text{ Hz})$; this phenomenon is known as aliasing, and it is a problem throughout frequency analysis. In general, to minimize the chance of aliasing, you should keep the spacing of your elements short enough such that all potentially influential frequencies are smaller than the Nyquist frequency. Otherwise, signals from these higher frequencies will be aliased into the other frequencies in your analysis, and you are likely to be fooled.

5.8.1   Fourier Algorithms

Fourier algorithms are a set of tools used to implement frequency analysis and are especially appropriate for investigating sinusoidal signals. You may be

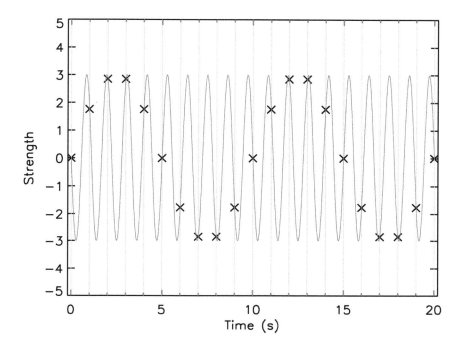

FIGURE 5.11
The simple wave pattern from Figure 5.10 is marked with x's. The times when
the function is sampled are marked with vertical light grey lines. A function
that could unexpectedly generate the observed pattern is drawn in dark grey.

familiar with Fourier expansions and transforms from mathematical studies
where you applied them to continuous functions.[10] The basic principles of
using these techniques on data sets are the same as the ones you may already
know from continuous functions. Before discussing the algorithms for real
data sets, let's briefly review the Fourier mathematics of continuous functions.
Readers wanting a more thorough discussion of the application to continuous
functions should read Chapter 10 of Kreyszig (1993).

Most continuous functions can be broken down into a sum of periodic
functions. The technique of using sine and cosine functions to perform this
decomposition is called a Fourier expansion. Other expansions using alternate
wave functions are possible, and may in fact be preferred when the periodic
oscillations in the continuous function do not have a sinusoidal shape.

---

[10]The use of the term "continuous" here is different than in Section 2.3, where it was
applied to element scales. Here, continuous refers to a smooth function, like $\psi = x^2$, without
accompanying measurements.

A Fourier transform is a technique used to change a continuous function $\psi(x)$ into a spectrum (also known as a spectral density) $F(\omega)$ – a spectrum has large amplitudes at ω that match the angular frequencies at which the continuous function oscillates. Because the spectra $F(\omega)$ are often complex (meaning they have both real and imaginary components), instead of directly examining the spectra it is useful to examine the relative power of $\psi(x)$ in different angular frequencies. If a function has features that repeat on some time or distance scale, they will appear as peaks in the power spectrum $P(\omega) = |F(\omega)|^2$. The total power contained in $\psi(x)$ can be calculated by integrating the power spectrum over ω. Through Parseval's theorem, the integrated power in Fourier space is always equal to the integrated power in real space, $\int |\psi(x)|^2 dx$.

What happens when you apply a Fourier algorithm to a continuous function that has oscillations that are not sinusoidal in shape? The frequency of the oscillations will still be detectable, but you will also get signal in other frequencies. A common example of this phenomenon is "ringing," which is produced when you take the Fourier transform of a function with a sharp edge (a decidedly not sinusoidal shape).

5.8.1.1   Fourier Transforms

How can Fourier algorithms be applied to a 1-dimensional data set to produce information regarding the data set's power in different frequencies? There are two groups of commonly used algorithms for performing Fourier transforms on real data: the Discrete Fourier Transform (DFT) and the Fast Fourier Transform (FFT). Both the DFT and the FFT can be used on any evenly spaced ratio-scale or interval-scale 1-dimensional data set, and they produce the same outputs and are interpreted in the same way. My discussion of these two techniques is heavily influenced by the more technical treatments in Press et al. (2007), Oppenheim et al. (1999), and Heiles (2005).

The DFT of an evenly spaced 1-dimensional data set is given by:[11]

$$Y_k = \sum_{j=0}^{n-1} \psi_j e^{\frac{-i2\pi k j}{n}} = \sum_{j=0}^{n-1} \psi_j e^{\frac{-i2\pi k (x_j - x_0)}{(x_{n-1} - x_0)}}, \qquad (5.24)$$

where $k = 0, 1, \ldots, n-1$. Be aware that the negative sign in the exponential is a convention that appears only in some versions of the DFT. The outputs of the DFT, the Y_k, are known as Fourier components; they are a series of complex numbers, each of which contains the amplitude and phase for a particular frequency of the DFT. The amplitude of each component is given by $|Y_k|$ and the phase by $\tan^{-1}(\mathbf{imag}\ Y_k / \mathbf{real}\ Y_k)$, where the **imag** and **real** functions return the imaginary and real parts of Y_k, respectively.

The units of the Y_k are rarely discussed – they are technically the same as the units of ψ_i. However, you can't make useful ratio comparisons between

---

[11]Throughout Section 5.8, the symbol i will denote imaginary numbers rather than an element index.

ψ_i and Y_k because the Y_k can grow with n, a number that can be arbitrarily large. Keep in mind that the units of the Y_k are different than the units of the continuous Fourier transform, which are those of the continuous function ψ multiplied by a factor of the units of the independent variable.

The far right-hand side of Equation 5.24 explicitly connects a Y_k component and a wave that completes k oscillations over the length of the data set. Euler's formula, given by $e^{i\omega\tau} = \cos\omega\tau + i\sin\omega\tau$, makes clear the connection between the exponent and sinusoidal waves with angular frequency ω. The $k = 0$ component, as you would expect, returns the sum of the ψ_k, also known as the <u>direct current</u> (DC) level of the signal. The relationship between the kth component of the transform and the frequency of the signal measured in physical units is given by $\nu_k = k/(x_{n-1} - x_0)$. However, to avoid breaking the Nyquist limit discussed in Section 5.8, something must happen for $k > n/2$, where $\nu_{n/2} = 1/2\Delta x$. It turns out that these larger values of k correspond to <u>negative frequencies</u>,[12] such that the full relationship of k to frequency is given by:

$$\nu_k = \begin{cases} \frac{k}{(x_{n-1}-x_0)}, & \text{for } 0 \leq k \leq \frac{n}{2} \\ -\frac{1}{2\Delta x} + \frac{k-n/2}{(x_{n-1}-x_0)}, & \text{for } \frac{n}{2} < k \leq n-1. \end{cases} \tag{5.25}$$

What does negative frequency mean? Mathematically, it is easy to state that functions with negative frequencies are written as $\sin(-\omega t)$ instead of $\sin(\omega t)$, but what does that mean physically? Basically, functions with negative frequencies have phases that evolve in the opposite direction as a function of the independent variable. Let's return to the conceptual picture of angular frequency as a point representing the phase spinning clockwise around a circle. The point could just as easily spin counterclockwise at the same rate – though the amplitude evolves similar to the clockwise case, this is a negative frequency. If you're only interested in the strength of oscillations of various frequencies, it's fair to analyze positive and negative frequencies together. You can do this by summing the absolute magnitudes, including both the real and imaginary components at the positive and negative frequency. Alternatively, you could use the power spectrum, discussed below.

A simple pseudocode algorithm for calculating the DFT is:

```
for k = 0 over k ≤ n − 1 {
        for j = 0 over j ≤ n − 1 {
              Y_k = Y_k + ψ_j exp(−i2πkj/n)
        }
}
```

This nested combination of **for** loops is the hallmark of an n^2 algorithm, meaning than of order n^2 simple operations are necessary to complete the calculation. The time required to run an algorithm scales linearly with the

---

[12]This derivation, which is discussed in Press et al. (2007), relies on the periodicity of Equation 5.24.

number of operations, as long as the run time is CPU dominated (as in not dominated by drive read/write time, etc.). I'll use this idea later to compare speeds with the FFT.

How are the outputs from a DFT interpreted? Let's start by reviewing a result from continuous Fourier transforms, and then examine how it applies to the DFT. The continuous Fourier transform of a sine wave with frequency ν_0 is given by:

$$F(\nu) = \int_{-\infty}^{\infty} \sin(2\pi\nu_0 t)e^{-i2\pi\nu t}\,dt = \frac{i}{2}\left[\delta_\infty(\nu - \nu_0) - \delta_\infty(\nu + \nu_0)\right], \quad (5.26)$$

where the δ function is defined as:

$$\delta_c(x) = \left\{ \begin{array}{ccc} c & : & x = 0 \\ 0 & : & x \neq 0 \end{array} \right. \quad (5.27)$$

This means that the Fourier transform of a continuous sine wave is given by two spikes in imaginary space, one positive and one negative, at the positive and negative frequencies of the sine wave. Similarly, the DFT of a sine wave data set with n elements is given by:

$$Y_{k,\sin} = \frac{in}{2}\left[\delta_1(\nu_k - \nu_0) - \delta_1(\nu_k + \nu_0)\right]. \quad (5.28)$$

This structure is very similar to the output of the continuous Fourier transform. In particular, notice the factor of n in front of the DFT; as n approaches ∞ and the data set gets longer and longer, the result from the DFT approaches the result from the continuous transform. The connection between the amplitude of the input sine wave and the outputs of the DFT is also apparent. A sine wave with $A = 1$ produces two delta function outputs with amplitude of $in/2$.

The DFT of a simple sine wave illustrates how to interpret the outputs from a general DFT. Essentially, a sinusoidal oscillation will produce peaks and valleys in real and imaginary space at the corresponding positive and negative frequency. Of course, actual data rarely produce outputs this easily interpreted; noise, ringing, aliasing, and multiple overlapping patterns often result in confusing DFTs.

In Figure 5.12 I've plotted the absolute magnitude of the DFT for the voltage data set as a function of frequency, calculated from both the imaginary and real parts of each Fourier component. The peaks of the DFT at $\nu = \pm 0.01$ Hz are exactly where I expected from my initial examination of the data set, but a second series of peaks also appear at $\nu = \pm 0.125$ Hz. The ratio of the amplitudes of the two signals is 3:1. Apparently the fluctuation in the peaks of the voltage that was apparent in Section 5.5.1 is the result of a second periodic signal in the data set with a smaller amplitude and a higher frequency. Indeed, I created the voltage data set by summing two sine waves plus two noise functions, the first with a normal distribution and the

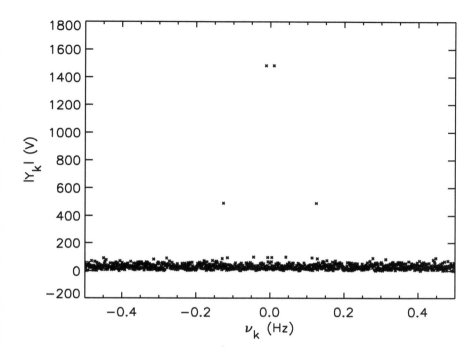

FIGURE 5.12
The absolute magnitude of the DFT components of the voltage data set, calculated from the real and imaginary components.

second with a uniform distribution. The appearance of two peaks in the DFT demonstrates the Fourier transform's linearity property; in other words, the Fourier transform of a sum of functions is equal to the sum of the individual DFTs.

Readers who are coding along with these examples will have noticed a subtlety in the analysis of the voltage data set: almost all of the power in the two signals detected comes from the imaginary part of the DFT. This happens because, as shown in Equation 5.28, the DFT of a simple sine wave is two imaginary delta functions.

One somewhat surprising characteristic of DFTs and FFTs is that they assume that outside the data set's range, the data repeat endlessly in a cycle. In other words, the algorithms operate assuming that after ψ_{n-1} at x_{n-1}, the value ψ_0 appears at $x_{n-1}+h$, followed by ψ_1 at $x_{n-1}+2h$, and so on. Similarly, the value ψ_{n-1} is assumed to appear at $x_0 - h$. This structure is identical to assuming that the data set is circular. To demonstrate this behavior, check the differences between the outputs from the DFT of ψ_i and those from ψ_i

padded with zeroes on both sides. Notice in the padded transform the ringing in the Y_k due to the sharp switching on and off of the data.

When a data set doesn't begin and end at the same phase of an oscillatory pattern, the fact that the Fourier transform assumes that the data repeat can cause signal in one frequency to leak into nearby frequencies because ψ_0 doesn't line up with ψ_{n-1}. In the voltage data set, this effect is actually what causes the data points near to $\nu = \pm 0.125$ Hz to have somewhat larger magnitudes than the background noise. I won't discuss how to minimize this effect in this book, but if you are interested to learn about it you should study data windowing (Press et al. 2007).

There are often occasions when you need to measure the strength of a signal at an arbitrary frequency ν which does not match with any of the ν_k given in Equation 5.25. In this case, you can use a slightly modified version of the DFT:

$$Y = \sum_{j=0}^{n-1} \psi_j e^{\frac{-i2\pi k_0 (x_j - x_0)}{x_{n-1} - x_0}}, \qquad (5.29)$$

where $k_0 = \nu_0(x_{n-1} - x_0)$, but keep in mind that if you just pick a bunch of frequencies out of a hat and evaluate them, your Y are likely to be correlated, which will make error analysis complicated and your total power incorrect. Better to use the standard ν_k when you can.

There are several techniques that allow you to measure the significance of your detected signals. One method that is computationally and intuitively straightforward (though not formally justified) is to compare the detected signals to those in a manufactured data set with random elements that have similar characteristics to the original data. First, measure the mean and standard deviation of your data set. Second, create a large number of fake data sets composed entirely of random numbers drawn from a normal distribution with the same characteristics. If you know that your data has a noise structure that is non-Gaussian, use that structure instead. The fake data sets should have the same spacing in x_i as the original data set. Finally, perform your Fourier analysis on all of the fake data sets, and measure the standard deviation of the measured signals (the mean should be nearly zero). These standard deviations represent metrics you can use to measure the significance level of the Fourier components detected in the real data set. Note that this method allows the significance level to vary as a function of frequency.

For reasons I will discuss shortly, there are times when you'll want to move from the Fourier transform of a data set back to the original 1-dimensional data set. To do this, use the <u>inverse DFT</u>, given by:

$$\psi_j = \frac{1}{n} \sum_{k=0}^{n-1} Y_k e^{i2\pi k j/n}. \qquad (5.30)$$

Equation 5.30 can be rewritten as $\psi_j = \mu + \sum_{k=1}^{n} Y_k \exp(i2\pi k j/n)$. This shows that you can recreate the data set ψ_j by summing the mean of the ψ_j plus a series of oscillating functions. The inverse DFT can be used as a convenient check on your code, for readers coding their own routines, because if you apply a Fourier transform to a data set, then an inverse DFT to those outputs, the original data set should emerge unaltered. Notice how similar the structure of the inverse DFT is to that of the DFT. Only the sign of the exponent and the normalization factor in front of the summation differ.[13]

The inverse DFT can also be used for interpolation. Call the value you are estimating $\widetilde{\psi}$, located at $\widetilde{x}$. First, perform the DFT on the data set to calculate the components Y_k. Then use those Y_k in the following formula:

$$\widetilde{\psi} = \frac{1}{n} \sum_{k=0}^{n-1} Y_k e^{\frac{i2\pi k(\widetilde{x}-x_0)}{(x_{n-1}-x_0)}}, \tag{5.31}$$

which is just the inverse DFT formula generalized for arbitrary x.

To keep the Y_k small, many authors will follow the convention of placing a normalization factor of $1/n$ in front of their DFT definition instead of in front of the inverse DFT. The issue is further confused by the variety of normalizations of the continuous Fourier transform pair, involving factors of 2π. Unfortunately, all you can do is be aware of this variety of normalizations and try to be consistent in your own work.

Fast Fourier Transforms

What about Fast Fourier Transforms? All FFTs work by breaking the ψ_i into smaller chains; this is known as a "divide and conquer" technique. The algorithm was popularized, though apparently not discovered, by Cooley and Tukey (1965). For n that are equal to powers of 2, the algorithm will proceed at $n \log_2 n$ speed, a great improvement over the n^2 speed of the DFT, especially when n is large. With some algorithmic cleverness the FFT does work for general values of n, thought the speed improvement may not be as prominent (it will certainly still be faster than the DFT). There are actually several different variants of the FFT, optimized for different values of n and internal computational operations.

The results output by the FFT are identical to those output by the DFT, so I won't review their interpretation. The only considerations you need to worry about are when to use the FFT instead of the DFT and how it works. My recommendation would be to use the FFT anytime the runtime of the DFT is affecting the rate that you can get work done, because there is certainly more algorithmic complexity to the FFT, and thus a greater likelihood of introducing bugs. If your analysis is fast enough as it is, keep it simple.

---

[13] The reason the normalization factor is necessary can be seen by examining the behavior of the Y_0 DC level.

5.8.1.2   Power Spectrum

As an alternative to directly plotting the outputs of the DFT, a plot of the power spectrum, P_k, can be used to examine the strength of the signal at different frequencies. When I use a Fourier transform, I usually don't need to view the positive and negative frequency components separately, so I prefer to analyze the power spectrum. The power spectrum's definition is given by:

$$P_k = \begin{cases} \frac{1}{n^2} |Y_0|^2 & \text{for } k = 0 \\ \frac{1}{n^2} \left(|Y_k|^2 + |Y_{n-k}|^2 \right) & \text{for } 0 < k < n/2 \\ \frac{1}{n^2} |Y_{n/2}|^2 & \text{for } k = n/2, \text{ even } n \text{ only} \end{cases} \tag{5.32}$$

which combines the power from positive and negative frequencies Y_k and Y_{n-k} into a single component of the power spectrum, P_k. The total power P in the signal can be found with the sum:

$$P = \sum_{k=0}^{n/2} P_k. \tag{5.33}$$

Since P_k is the sum of squared magnitudes, $P_k \geq 0$, and thus the same is true for P. The units of the power spectrum are those of the DFT, squared.

There is a lot of confusion in the literature regarding the form of the power spectrum because there are many different meanings for the term "power spectrum" that are related, but not identical, and are used interchangeably. Let's examine the example of our simple cosine wave, $A \cos(2\pi\nu x_i + \phi_0)$; what is the most intuitive value for the power spectrum to return? Personally, I would like it to output the average value of the function squared, $A^2/2$, but many analysts prefer other quantities so they've defined their normalizations and formulae differently. There are two main sources of normalization confusion. First, should the power spectrum be defined separately for both positive and negative frequencies, or should they be combined? Second, should the total power be equal to the mean squared amplitude of the function, or should the length of the signal influence the total power? Each of these choices is a convention, so it's important to be consistent. In this text, I'll combine positive and negative frequencies in my power spectra, and set the sum of the P_k to be the mean squared amplitude. For a thorough discussion of the multitude of possible normalizations, see Press et al. (2007).

Parseval's Theorem, which is also extremely useful in continuous transforms, can be used for real data as a check on your algorithms:

$$\sum_{i=0}^{n-1} |\psi_i|^2 = \frac{1}{n} \sum_{k=0}^{n-1} |Y_k|^2 = nP. \tag{5.34}$$

Note here how important it is to choose the correct frequency spacing; using some other pattern would result in a sum over an improper set of numbers and an incorrect P. This would occur because some portions of the spectrum would be undersampled and some would be oversampled.

In the voltage data set, there are two major contributors to the power spectrum. These are the two delta functions, one at positive frequency and one at negative frequency, from each of the two sine waves. The total power (P) in the signal is 6210 V^2, which can be calculated using either form of Equation 5.34.

5.8.2 Lomb Periodogram

The Lomb periodogram is a technique for finding waves and other oscillations in data sets with gaps or unevenly spaced elements (Press et al. 2007). It outputs the same amplitude you would find using a linear least-squares fit of a sinusoid (see Section 5.6), and also allows for a straightforward error analysis. In a Lomb periodogram, the data are weighted equally per element, as opposed to equally per unit in x.

Following Press et al. (2007), the power of a data set at frequency ν is given by (Lomb 1976; Scargle 1982):

$$
P(\nu) = \frac{1}{2}\left[\frac{\left(\sum_{i=0}^{n-1}(\psi_i - \mu)\cos(2\pi\nu(x_i - \tau))\right)^2}{\sum_{i=0}^{n-1}\cos^2(2\pi\nu(x_i - \tau))}\right.
$$
$$
\left. + \frac{\left(\sum_{i=0}^{n-1}(\psi_i - \mu)\sin(2\pi\nu(x_i - \tau))\right)^2}{\sum_{i=0}^{n-1}\sin^2(2\pi\nu(x_i - \tau))}\right], \tag{5.35}
$$

where τ is defined by:

$$
\tau = \frac{1}{4\pi\nu}\tan^{-1}\left[\frac{\sum_{i=0}^{n-1}\sin(4\pi\nu x_i)}{\sum_{i=0}^{n-1}\cos(4\pi\nu x_i)}\right]. \tag{5.36}
$$

The units of the Lomb periodogram are power: the amplitude units of ψ squared, just like in a power spectrum calculated from the DFT. Notice the two sums in the definition of τ in Equation 5.36 depend solely on the measurement locations x_i and not the elements ψ_i, so if you have multiple data sets with the same spacing in x, then τ only has to be calculated once per frequency. Calculating a single $P(\nu)$ requires of order n operations, so the full set of $P(\nu)$ requires of order mn, where m is the number of frequencies evaluated. You are basically free to evaluate whichever frequencies you want, with the warning that the ones you choose are unlikely to be independent.[14]

Let's take a quick look at the structure of Equation 5.35: it's the sum of two terms, one dependent on sine functions and the other dependent on cosines. You can think of the terms in the denominators as similar to a normalization

---

[14]A faster and more elegant algorithm structure is discussed in Press et al. (2007), but be careful as their version of the Lomb periodogram is normalized by a factor of σ_{n-1}^2.

based on the structure of x, since they don't depend on the ψ_i. The numerators multiply the data by simple sine and cosine waves of the chosen frequency and sum the response.

Now I'll step through the Lomb periodogram of a simple sine function on a data set with evenly spaced elements to demonstrate how the algorithm works. Following Equation 5.35, only the second term in brackets is going to contribute a significant amount, because we know from calculus that the integral of a sine function times a cosine function over some interval is zero. Let's make a simplifying assumption that the data set has an element spacing where τ is zero, and since this is a simple sine function, μ is also zero. Next, in the numerator, two sine functions are multiplied together. When the frequencies of the sine waves match, the functions will constructively multiply and the sum will be nonzero. When they don't match, the signals will interfere and the sum will be close to zero. Then, the only power remaining in the Lomb periodogram will be at the frequency of the input signal.

Of course, when the data set contains unevenly spaced or missing elements, things are not so easy.[15] One way to see the complexities resulting from unevenly spaced elements is to take a Lomb periodogram of a simple sine wave measured on an uneven grid of elements; some of the wave's power will leak into other frequencies. This demonstrates that even if your data only contains signal at one frequency, there will be power in other frequencies, though in most cases it will not be at a significant level.

Comparing the power output by the Lomb periodogram with the amplitude of the signal input is not straightforward. If you apply the Lomb periodogram to evenly spaced data, you can "renormalize" Equation 5.35 by multiplying by $2/n$, but the renormalization won't work for unevenly spaced data and the Fourier transform is a better technique to apply to evenly spaced elements. The DFT and the FFT are much more versatile, mostly due to their ability to reproduce the ψ_i from the Y_k. Since the frequencies measured by the Lomb periodogram are not independent, you can't use it to reproduce the ψ_i.

The random data set generation mechanism described in the previous section is appropriate to quantify the significance of the signals detected with a Lomb periodogram. Again, you'll want to choose the noise model carefully, based on the structure of your data; don't just choose normally distributed noise because it's easy.

Another consequence of using the Lomb periodogram on unevenly spaced data is that it is possible to be fooled by a false signal originating due to ringing from interference between different frequency components, particularly when there are a handful of modes that are relatively powerful. As always, the best check is to visually inspect the data; if the periodogram indicates a strong signal you can't see any evidence for, be skeptical.

The temperature data set is an excellent place to test the Lomb peri-

---

[15]Mathematically, this occurs because simple sine and cosine waves are not eigenmodes of the system.

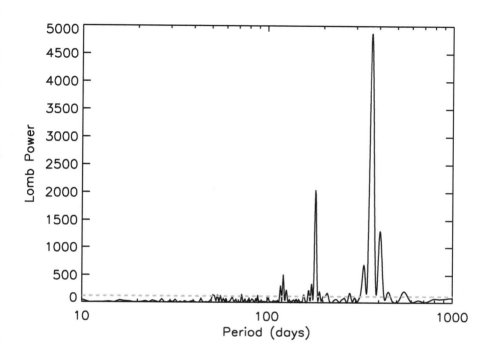

FIGURE 5.13
Power in a Lomb periodogram analysis on the temperature data set. The 99%
significance level, based on the model of normally distributed noise, is shown
by the dashed line. The largest peak is at a period of 365 days, and the second
is at 182 days. Note the logarithmically scaled x-axis.

odogram. The results of that analysis are shown in Figure 5.13. I've evaluated
the periodogram at a dense set of oversampled periods in between 10 and
1000 days. The peak response appears at frequency $\nu = (1/365)$ days^{-1}, corresponding to the annual temperature cycle. There are several other peaks in
the data that cross the 99% significance threshold, including one at a period
of 182 days. This peak, since it occurs at a frequency directly related to the
largest signal in the data, should be treated with skepticism; often these are
artifacts caused by ringing. However, it turns out to be real! It seems Ethiopia
has a rainy season from July to September, which lowers the temperature during those months. The highest temperatures are before and after this period.

Is there a way to confirm this strange behavior? Of course – take the
mean of the average temperatures for each day of the year, e.g., average all
the January 1 elements together (Figure 5.14). To keep things simple, I've
ignored leap days. This clearly demonstrates the double-peaked nature of the
yearly temperature cycle.

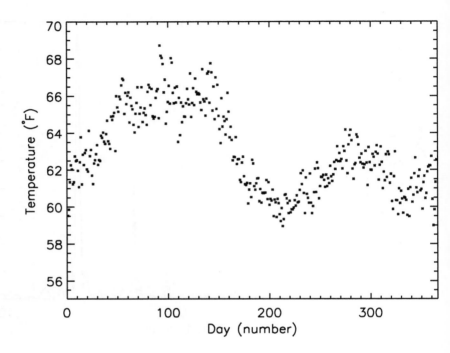

FIGURE 5.14
Mean of the average temperatures for each day of the year in the temperature data set.

5.9 Filtering

There are many situations in which you want to minimize the effects of a particular range of frequencies on a data set. Building on the frequency analyses of Section 5.8, filtering is a powerful technique for manipulating which frequencies your analysis is sensitive to. Imagine a data set contaminated with a 60 Hz signal from the electrical power used to run the instrument. Wouldn't it be nice to eliminate the influence of that frequency while leaving the rest of the data unaffected? What about a stock that suffers from a large seasonal variation. How can you eliminate the year-period oscillation to see other trends? Filtering will allow you to meet these challenges.

For data sets with evenly spaced elements, filtering in the Fourier domain is far and away easier than doing it in the x domain. The algorithm for the technique is quite straightforward. First, take the DFT or the FFT of your data. Next apply the filter to the Y_k to create the filtered $\widehat{Y_k}$, and then pass $\widehat{Y_k}$ through the inverse DFT to construct the filtered $\widehat{\psi_i}$.

There are many types of filters that could be applied to your data. In fact, you have great freedom to tune your filter to pass whichever frequencies you like, but there is no reason to filter ν outside the range of valid frequencies from Equation 5.22. These frequencies don't appear in your Y_k anyway (except through aliasing to frequencies within the range of response). It's generally better to build a filter based on wavelength or frequency than wave number, because λ and ν are quantities that are directly related to a physical scale.

High-pass filters allow high frequencies through while damping low frequency ones. An example of a high-pass filter is given by:

$$\widehat{Y_k} = \begin{cases} 0, & \text{for } \frac{1}{|x_{n-1}-x_0|} \leq |\nu_k| \leq \nu_c \\ Y_k, & \text{for } \nu_c \leq |\nu_k| \leq \frac{1}{2\Delta x}, \end{cases} \tag{5.37}$$

where ν_c is a constant that sets the scale of the filter and is chosen by the analyst. Notice the filter depends on the magnitude of ν_k, so that positive and negative frequencies are treated identically. This particular filter is sharp, in that the Y_k are either passed untouched or zeroed out. It is also possible to construct filters with smoother effects in frequency.

From Equation 5.37, it is easy to see how to construct a low-pass filter, which has the same structure, but reversed criteria. Bandpass filters allow a range of frequencies through, with high and low frequency limits. Notch filters eliminate a single frequency, while leaving the other frequencies unaffected. Note that the smoothing techniques discussed in Section 5.5 are actually forms of low-pass filters, which happen to be convenient to calculate in x-space.

In Section 5.8 I demonstrated that the DFT of the voltage data set consists of noise plus signal at two frequencies. Let's pretend that the signal at $\nu = \pm 0.125$ Hz is an artificial contaminant. Figure 5.15 shows the results when a notch filter is applied at that frequency to remove the contaminant. Notice how the fluctuations in the oscillation peaks that were evident in Figure 5.2 have been removed by the notch filter; now the signal is just a simple sine wave plus noise.

5.10 Wavelet Analysis (Advanced Topic)

Wavelet analysis is a modern computational technique used to provide simultaneous position and frequency analysis.[16] In contrast, Fourier and other frequency analyses can tell you about the strength of different frequencies, but

---

[16] In this section, I have relied heavily on two excellent and accessible introductory guides to wavelets: Torrence and Compo (1998) and Rosolowsky (2005).

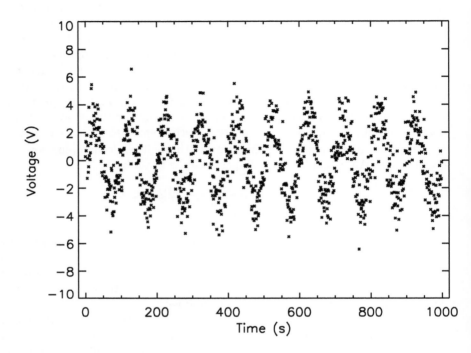

FIGURE 5.15
The voltage data set with the $\nu = \pm 0.125$ Hz signal removed via a notch filter.

not about where in the data set a given frequency is strong. A word of caution regarding wavelets: many decision makers and leaders have no experience interpreting them, so additional care is necessary to visualize and communicate findings. The situation is even worse than for frequency analyses, since at least those techniques are well-established and frequency is a popularly known concept. Wavelets are completely foreign to most people.

Wavelet analyses can be even more powerful when used in conjunction with frequency analyses. For example, I was once analyzing a circular data set with two large gaps on opposite sides of the circle. I was interested in detecting oscillations that did not appear over the full 2π around the circle. Since the gaps made the data set unevenly spaced, I used a Lomb periodogram to determine which frequency oscillations were present and then a wavelet transform to find the local regions where these oscillations were strong. Using only the Lomb periodogram would not have given me positional information, and using only the wavelet transform would have made it difficult to tell which frequency perturbations were the most powerful throughout the data set.

The basic idea of wavelet analysis is to cycle through the positions x_i in the data set and examine the response to different frequency oscillations at each point. This can be broken down into an algorithm with a few steps:

1. Choosing a mother wavelet,

2. Calculating the wavelet transform, and

3. Analyzing the outputs of the wavelet transform.

5.10.1 Choosing a Mother Wavelet

The first step in performing a wavelet analysis is to choose a <u>mother wavelet</u>, also known as a mother function, which I will represent by $M(\kappa)$. The mother function in a wavelet analysis is the shape that is compared to the signals in the data set. To get the best results from your wavelet analysis, choose a mother function that is similar to the shape of the oscillations in the data. For example, if you see oscillations that look sinusoidal, you should choose a mother wavelet that includes a sinusoidal function. The same holds true for sawtooth or boxcar shapes. This is similar to the choice of sinusoidal functions from the earlier discussion of frequency analysis.

Mother wavelets have some standard characteristics. First, the integral of $M(\kappa)$ over all κ must approach 0. Secondly, the function must be square integrable, meaning:

$$\int |M(\kappa)|^2 d\kappa = \text{Some finite number.} \tag{5.38}$$

Functions that meet this criteria are well-behaved (i.e., they don't go running off to ∞, etc.). In practice, I will normalize $M(\kappa)$ such that "some finite number" is 1. I do this so that mother functions that differ only in the value of a parameter contribute the same power, and are thus directly comparable. Unnormalized wavelets will be marked with the subscript 0.

One of the most commonly used mother wavelets is the Morlet wavelet (shown in Figure 5.16a), which takes the form:

$$M_{\text{Mor},0}(\kappa) = \pi^{-1/4} e^{i\omega\kappa} e^{-\kappa^2/2}, \tag{5.39}$$

where ω is the angular frequency of oscillation. The Morlet wavelet is assembled by multiplying a sinusoid times a Gaussian. In this parameterization, the Gaussian always has a standard deviation of 1, so the choice of the parameter ω sets the number of oscillations completed before the falling amplitude of the Gaussian damps the amplitude to near 0. If you would like n oscillations to be completed within $-1 \leq \tau \leq 1$, choose $\omega = n\pi$.

A similarly shaped mother wavelet can be constructed from the derivatives of a Gaussian (DOG). It has the general form:

$$M_{\text{DOG},0}(\kappa) = \frac{(-1)^{c+1}}{\sqrt{\Gamma(c + \frac{1}{2})}} \frac{d^c}{d\kappa^c} e^{-\kappa^2/2}, \tag{5.40}$$

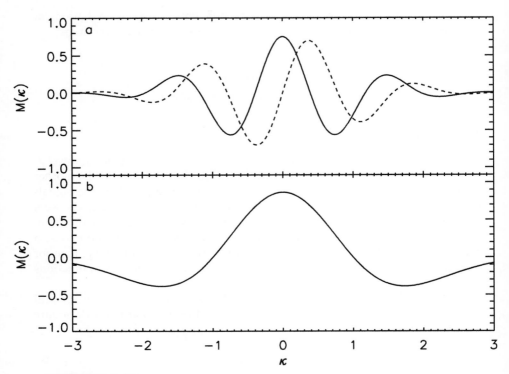

FIGURE 5.16
a: The Morlet mother wavelet for $\omega = 4$. The real part of the function is drawn with a solid line, and the imaginary part with a dashed line. b: The Derivative of Gaussian mother wavelet for $c = 2$ (sombrero).

where c is the number of the derivative, and the Γ function is a normalization factor derived from the factorial function (see Kreyszig (1993)). No common convention is present in the literature for the normalization of this wavelet. The $c = 2$ "sombrero" version of this mother wavelet is used frequently (Figure 5.16b):

$$M_{\text{DOG2,0}}(\kappa) = \frac{2}{3^{1/2}\pi^{1/4}}(\kappa^2 - 1)e^{-\kappa^2/2}, \qquad (5.41)$$

where I have used the relation $\Gamma(5/2) = 3\sqrt{\pi}/4$.

A major difference between wavelets and Fourier transforms is that, unlike sine and cosine waves of different frequencies, the multiple of two different frequency mother wavelets does not integrate to zero. In mathematical terms, this means that the functions are not orthogonal, and as a result the transforms at different scales and positions are not independent.

5.10.2   The Wavelet Transform

After choosing a mother wavelet, the next step is to compare the $M(\kappa)$ to the data using the <u>wavelet transform</u>. Unlike Fourier transforms, which only vary the frequency of the response function, wavelet transforms calculate the response of the signal to a mother function by varying over a 2-dimensional space, parameterized by both frequency and position. This extra information doesn't come for free; you pay for having information in two dimensions by sacrificing precision in both.[17] The mother wavelets we described in the previous section are all functions of the variable κ. To be able to shift the $M(\kappa)$ in position and frequency, all I have to do is make the substitution: $\kappa = \frac{x-a}{b}$. Then, changing a shifts the position of M, and changing b varies its scale. Note that in this context, scale refers to the frequency of the oscillations and the width of the mother wavelet, but not the amplitude of $M(\kappa)$. To translate this parameterization into a form that can be used with real data, use the substitution:

$$\kappa = \frac{x_i - a}{b} = \frac{(i-j)\Delta x + x_0}{b}, \tag{5.42}$$

where $a = j\Delta x$. With this parameterization, I normalized $M_0(\kappa)$ with the relation:

$$M(\kappa) = \frac{1}{\sqrt{b}} M_0(\kappa) \tag{5.43}$$

The function stretches and compresses as the scale changes, but the normalization factor keeps the area under the square of the function constant.

At this point, I can connect the oscillations of the mother wavelets to the physical frequencies to which they are responsive. For the Morlet wavelet, this is slightly different from the frequency of its sinusoidal component due to the influence of the Gaussian. The relation between the wavelength of the oscillation probed, λ, and the angular frequency input to the wavelet is given by:

$$\lambda_{\text{Mor}} = \frac{4\pi b}{\omega_0 + \sqrt{2 + \omega_0^2}}. \tag{5.44}$$

This is very close to the usual relationship between wavelength and angular frequency, $\lambda = 2\pi/\omega_0$, multiplied by the scale factor b, which would have been an intuitive guess. For the derivative of a Gaussian mother wavelet, no such guess is possible; the relation is:

$$\lambda_{\text{DOG}} = \frac{2\pi b}{\sqrt{c + \frac{1}{2}}}. \tag{5.45}$$

The wavelet transform of an evenly spaced data set with elements ψ_i is given by:

$$W_{j,b} = \sum_{i=0}^{n-1} \psi_i M^* \left[\frac{(i-j)\Delta x + x_0}{b} \right] \tag{5.46}$$

---

[17]Physicists will recognize this as a form of an uncertainty principle.

where n is the number of elements in the data set, i is a position index, Δx is the spacing of the elements in x-space, and M^* is the complex conjugate of the normalized mother wavelet (Torrence and Compo 1998). Without orthogonality, the wavelet transform can be performed on an arbitrary series of scales, though they won't be independent. It is often convenient to have the resolution in scale space vary exponentially, such as:

$$b_k = b_0 2^{k\Delta k}, \quad k = 0, 1, \ldots, K, \tag{5.47}$$

where b_0 is the smallest scale that can be sampled, $2\Delta x$, analogous to the Nyquist limit for frequency analysis discussed in Section 5.8. The quantity Δk is a measure of how dense the sampling in scale space is, and it must be small enough so that none of the signal slips between the cracks. Using $\Delta k = 0.5$ will generally be sufficient, and you can use a smaller scale if you have the computational cycles to burn!

A wavelet transform can be a computationally intensive procedure. Consider that you will be evaluating m frequencies at n locations in the data, and each evaluation takes n operations. Therefore, a full wavelet transform takes something like mn^2 operations. If you try to apply the full set of transforms without thinking about what you are doing, you're liable to sit in front of your computer for a long time. So, if you're not interested in the full range of frequencies, restrict their range. If you don't need the small spacings in your data set, interpolate onto a coarser grid, and don't evaluate the high frequency parts of the transform. Of course, if you have sufficient computational resources, you don't have to worry about any of this.

Related to the physical scale of the wavelet is a concept called the cone of influence. I discussed a similar idea, the zone of influence, in the development of the running mean in Section 5.5.1. Basically, if your wavelet is calculated centered on a position close to the edge of the data set or a gap, the $W_{j,b}$ will be biased by the lack of data over part of its range. This bias stretches further into the data for larger wavelength scales, since they respond over larger ranges of x. The biased parts of position-scale space are within the cone of influence of the edges. It's cone shaped because it is wider at larger scales than it is at smaller ones.

Quantitatively, the distance the cone of influence extends into the data is given by $x_{\text{cone}} = \sqrt{2}b$ for both the Morlet wavelet and the Derivative of Gaussian (this is not the solution for all wavelets). These distances correspond to the place where the effect of the edge falls off by a factor of e^2 in power.

From Equation 5.46, I construct the wavelet power spectrum, $P_{j,b} = |W_{j,b}|^2$. Note that this is a similar relation as for the Fourier transform, so the intuition you've developed for Fourier power spectra also applies here.

5.10.3   Analyzing the Outputs of a Wavelet Transform

Now that I've defined the wavelet power spectrum, it would be helpful to have a measure of the significance of the signals detected. Let's discuss how to compare our signals to Gaussian distributed noise, but keep in mind, if your noise model isn't well described by Gaussian noise, don't use this measure of significance! Torrence and Compo (1998) contains an excellent discussion of types of red noise, amongst other topics. Using a model of Gaussian noise with the same standard deviation as the data set, the 95% confidence level for power (see Section 3.7) is given by:

$$P_{95\%} = \begin{cases} \frac{3.84146}{2}\sigma_n^2, & \text{for Morlet} \\ 3.84146\sigma_n^2, & \text{for DOG.} \end{cases} \tag{5.48}$$

The constant in front of the both relations comes from integrating over a χ^2 distribution, a topic that I will not describe in this book. The factor of two difference between the significance level results from the Morlet wavelet having twice as many degrees of freedom as the DOG, due to its real and imaginary parts.

The middle panel of Figure 5.17 shows the wavelet power spectrum of the elevation data set, with the 95% confidence limit marked. For this analysis, I chose the DOG wavelet with $c = 2$, which, to my eye, looks somewhat similar to the shape of the continents above water. Before transforming the data, I interpolated it onto a grid with spacing $\Delta x = 0.5°$ to keep the computational time manageable. The scales calculated are calculated from Equation 5.47 with parameters $b_0 = 1°$, $\Delta k = 0.5$, and $K = 11$. For readers coding these examples, make sure you calculate the standard deviation of the interpolated data set over 360°, not 540°; the hemisphere is repeated only for ease of reading the plot!

There are strong responses located at each peak, as well as each valley, at a scale of $\sim 100°$. The valleys also produce a strong signal because they are responding to the negation of the mother wavelet. There are sharp falloffs in the signal strength at the longitudes where the continents rise out of the water, because at these places the shape of the data is nothing like the mother wavelet. Note that the signal peak near $-135°$, corresponding to the Pacific Ocean, is located at a larger period than the other peaks. This occurs because the Pacific is the longest feature in the data. Also note that there is no cone of influence for this data set, since it is circular.

The bottom panel also shows a wavelet power spectrum, this time calculated with the Morlet wavelet with $\omega_0 = 6$. As the wavelengths probed by the two mother functions are slightly different, for this analysis I used $K = 13$ instead of the 11 I used for the DOG function. Since the Morlet mother function includes several oscillations, instead of just a declining peak, the strongest signal is found near 45°, again with a wavelength near 100°. Note that this wavelength does not correspond to the full range of the mother wavelet, since it contains several oscillations. Rather, it is the peak to peak distance.

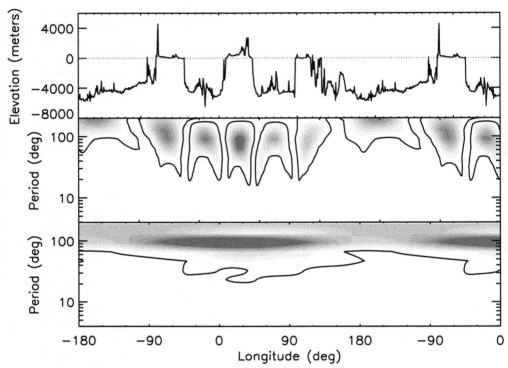

FIGURE 5.17
The top panel shows the elevation data set after it has been interpolated onto a 0.5° grid, with a dashed line marking sea level. The middle panel shows the wavelet power spectrum calculated with the DOG mother function with $c = 2$, the solid line representing the 95% confidence level for power, and each shaded contour representing 5σ detection, relative to Gaussian noise. The bottom panel is the power spectrum calculated with the Morlet wavelet with $\omega_0 = 6$.

5.11 Exercises

1. Construct a histogram of the points in the elevation data set, demonstrating that the distribution of elements is bimodal.

2. In the temperature data set, the measurements for March 1-4, 2000, are missing. Use linear interpolation on the adjacent elements to derive estimates for the missing data. Do you feel like these are accurate estimates? Why or why not?

3. Plot three versions of the elevation data set on the same set of axes: the original data set, the mean smoothed data set, and the median

smoothed data set. Set $w_h = 200$. Compare and contrast the three versions.

4. (Advanced) Calculate the running mean for the temperature data set using $w_h = 5$. Exclude elements that would be influenced by edge effects.

5. Perform a straight-line fit to the age list between the ages of 0 and 44 years. What are the best estimates, along with the uncertainty, in the fitted parameters?

6. (Advanced) Confirm the fits and uncertainties that were calculated for the age list with the straight-line fit formulae using the linear algebra formulation.

7. Confirm Equation 5.28 by using the DFT algorithm on a sinusoidal data set you construct. Then, show the linearity of the Fourier transforms by constructing a data set that is the sum of two sine waves of different frequency and using the DFT on it.

8. Demonstrate the phenomenon of ringing by calculating the Fourier transform of a evenly-spaced 1-dimensional circular data set with $n = 360$. The data set is defined as follows: for odd degrees, $\psi = -1$. For even degrees, $\psi = 1$.

9. (Advanced) Calculate and plot the wavelet power spectrum for the voltage data set using a Morlet mother function. Include the 95% confidence level for signal and mark the cone of influence of the boundaries. Can you explain the peaks in the signal?

6

Related Lists and 1-Dimensional Data Sets

6.1 Introduction

Oftentimes, instead of having a single list or 1-dimensional data set to analyze, an analyst will have multiple related data sets to track and compare. For example, imagine you had a 1-dimensional data set of the daily high temperatures in Chicago along with the number of incidents of heatstroke reported at area hospitals. Or, perhaps you have the prices of two related stocks, both of which vary as a function of time. It is useful to compare these data sets and to examine whether variations in one are connected to variations in the other. Comparisons can illuminate links between the data sets, causal or otherwise, and can confirm or contradict your understanding of the system of interest. The techniques I will discuss in this chapter are flexible enough to apply to a variety of situations and the analyst's imagination is the only limit when it comes to designing comparisons.

Many of the techniques discussed in this chapter can be used when the data are stored in two lists (see Chapter 3) with related pairs of elements across the lists. For example, imagine the first list contained the heights of buildings across New York City, and the second list contained the amount of square footage in each building. The two lists have related pairs, since each building can be associated with a height and a square footage.

When the related data structures are 1-dimensional data sets, the associated location information allows the application of additional techniques. To be amenable to these analyses the 1-dimensional data sets should be functions of the same location variable, but the elements can be measured at completely overlapping times (or positions) or with no overlap whatsoever. The elements of the two data structures often have the same units, but they also can be dimensionally unrelated. However, the elements of each individual list must still be internally consistent.

In this chapter, I will use a formalism that can accommodate both lists and 1-dimensional data sets and is similar to the notation I introduced in the previous chapter. Let's call the first data structure ψ_i and the second ϕ_j. If the data structures are 1-dimensional data sets, they are measured at locations x_i and y_j. Note that ψ and ϕ are not necessarily measured with the same units, but x and y must be. The data structures ψ and ϕ can have an unequal number of elements. If the data structures are 1-dimensional data sets, x and

y can have different spacings. In some cases, the two location arrays x and y may not even overlap; for example, if ψ and ϕ counted the number of traffic tickets issued under different policies, x could span the month of August and y the month of October.

6.2　Jointly Measured Data Sets

Though the formalism can accept data structures that have differing numbers of elements or are measured at different locations, comparing related data sets is most straightforward when they have the same number of elements and are measured at the same locations. Sometimes, structures meeting these qualifications, which I will call jointly measured data sets, can be constructed by extracting elements from the original data sets. To distinguish the jointly measured data sets from the original data, I will mark them with an additional subscript z: $\psi_{z,i}$ and $\phi_{z,i}$. For 1-dimensional data sets, let z_i represent the locations where both ψ and ϕ have been measured (no parallel location array is necessary for lists). Additionally, the two jointly measured data structures must have the same number of elements m.

The degree of manipulation of the original data necessary to create the jointly measured data sets depends on the state of the original data. Potentially, these jointly measured data sets are subsets of the measurements in ψ and ϕ, because they don't include ϕ at locations where ψ has not been measured, nor do they include ψ at locations where ϕ has not been measured. Some data sets may require no modification whatsoever, but for others, especially 1-dimensional data sets without structure in x or y, it may not even be possible to construct jointly measured arrays. Be careful that you do not introduce bias through this manipulation; this could occur if the excluded elements were, for whatever reason, not random.

Let's step through a simple example of creating jointly measured data structures. Imagine you had the following 1-dimensional data sets:

$$\psi = [5, 3, 8, 0, 2, 6] \text{ measured at } x = [0, 1, 2, 3, 4, 5]$$
$$\phi = [1, 3, 5, 0, 9, 8] \text{ measured at } y = [0, 2, 4, 6, 8, 9] \tag{6.1}$$

Then, the jointly measured data structures are given by:

$$\psi_z = [5, 8, 2] \text{ and } \phi_z = [1, 3, 5] \text{ measured at } z = [0, 2, 4] \tag{6.2}$$

In other words, just look for the locations where ψ and ϕ are both measured, and exclude locations where only one or none are measured.

For related pairs of lists, similar manipulations may be required. For example, imagine that for the square footage and heights of buildings lists, some buildings lacked measured heights and others lacked measured square footages. In this case, buildings that didn't have measurements for both quantities would have to be eliminated from the jointly measured lists. As usual, when this happens it is up to the analyst to determine whether excluding some of the data from the jointly measured data sets introduces biases. This kind of bias would occur, for example, if the person taking square footage measurements was unable to complete his work for buildings taller than 10 stories.

There is one special case where 1-dimensional data sets with non-overlapping x and y can be manipulated to form a useful z: when x and y have a constant offset. Then, it is possible to create the array z by adding or subtracting a constant factor from x or y. For the example of daily traffic ticket counts in two different months, this can be accomplished by subtracting 2 months from the October dates so that x and y overlap. When this technique is used, it is imperative to make clear when presenting all of the analyses that the data were manipulated in this way, because the offset often has implications for the model.

6.3 Example 1-Dimensional Linked Data Sets

In this chapter, I will augment three data sets from earlier chapters with additional data so I can apply new analytical techniques.

6.3.1 Weights of People in a Classroom

In Section 3.2.1, I introduced a list of the heights, in inches, of 33 occupants of a middle school classroom. Let's supplement that list with the weights of these same occupants. This creates a related pair of lists; each height is associated with a weight. As long as those associations can be preserved, mixing the order of the elements in both lists does not create a different data set. The height and weight pairs for this data set are shown in Table 6.1.

In reality, there are complex relationships that relate height and weight, but for this hypothetical data set I derived the weights, in pounds, by multiplying the heights in inches by two. I then added or subtracted a random number of pounds distributed normally, centered on zero, with a standard deviation of 10 lbs.

6.3.2 A Second Elevation Data Set

In Section 5.3.2, I introduced an evenly spaced circular 1-dimensional data set of elevations along the equator drawn from the ETOPO1 satellite maps. Here, I will add to that data set a second structure with elevations measured just 5" north of the equator. In linear distance, this corresponds to approximately

TABLE 6.1
Heights and weights of the 33 occupants of a middle schools classroom. Heights are in inches and weights are in pounds.

Height	Weight	Height	Weight	Height	Weight
46.9	85.3	50.7	87.8	51.8	100.7
48.5	95.3	50.8	102.1	52.2	109.6
48.6	99.0	50.8	111.8	52.3	82.2
49.0	114.1	51.0	89.8	52.4	107.4
49.1	96.4	51.1	109.4	52.6	104.4
49.1	104.8	51.2	111.1	52.9	89.3
49.7	93.9	51.3	94.7	53.1	109.1
50.1	102.2	51.5	106.3	55.2	113.5
50.4	110.1	51.6	105.2	66.2	143.3
50.5	113.0	51.6	101.6	67.8	143.0
50.5	116.4	51.7	100.2	69.7	137.8

6 miles. This second 1-dimensional data set has the same number of points as the equatorial one, 21600, but the points are located ever so slightly closer together due to the path around the globe being very slightly shorter at 5" north of the equator. The highest elevation measured on this northern strip is 4029 m, the lowest is −7160 m, the mean is −2901 m, the median is −3823 m, and the standard deviation is 2124 m. Just 21% of the points are located above sea level.

The features of this new data set are similar to those of the data set measured at the equator. The difference between the two elevation data sets is plotted in Figure 6.1, in grey. The similarity between the two data sets is obvious at first glance; this is as expected, since I would be surprised if the elevations above sea level of two locations separated by only 6 miles were not similar! To the naked eye, the difference between the two data sets appears to be centered around zero (though that did not necessarily have to be the case), and has spikes that can reach approximately 2000 meters in magnitude.

6.3.3 A Second Temperature Data Set

In Section 5.3.3, I introduced a 1-dimensional data set of the average daily temperature in Addis Ababa measured at evenly spaced elements with gaps. Here, I introduce a similar data set of temperature measurements for Nairobi in Kenya (Figure 6.2), a city located about 1000 miles to the south of Addis Ababa. This data set spans from 1 January 1995 to 24 December 2006 for a total of 4376 entries. Of these, there are data recorded for 2847 days. There is also a large gap in this data set, and the recordings end about 2 years earlier than the Addis Ababa data. The maximum average daily temperature is

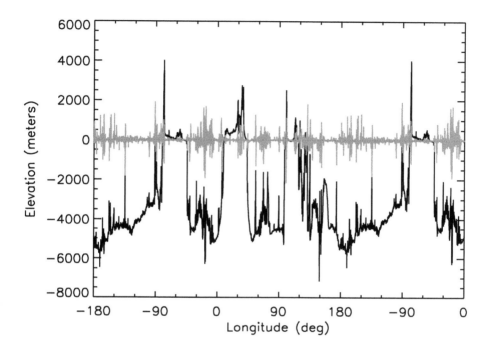

FIGURE 6.1
Elevation as a function of longitude 5" north of the equator is drawn with a heavy black line. The solid grey line is the difference between the elevations in the equatorial data set and the one 5" north; positive values correspond to longitudes where the equatorial data set is higher. Negative longitudes are in the Western hemisphere, and positive in the Eastern.

82.4°F and the minimum is 51.8°F. For elements excluding the gaps, the mean is 67.3°F, the median is 67.2°F, and the standard deviation is 4.0°F. Speaking generally, these statistics appear about 5° warmer than their equivalents in the Addis Ababa data set. An annual temperature cycle for Nairobi similar to that seen for Addis Ababa in Figure 5.14 is also readily apparent (Figure 6.3), but displaced about 5° warmer.

6.4 Correlation

Two data sets have a <u>correlation</u> if the value of one is connected to the value of the other in a statistically significant way. In other words, when two data sets

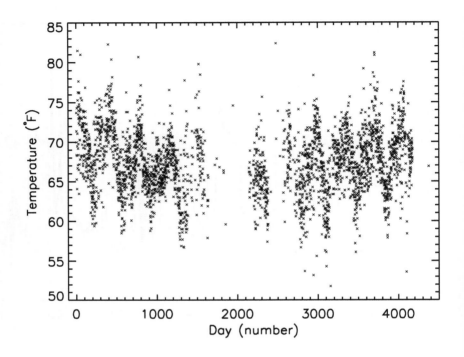

FIGURE 6.2
Average daily temperature in Nairobi, Kenya, from 1 January 1995 to 24
December 2006.

are correlated knowing the value of one data set gives you some predictive
power over the other. For example, height and weight are correlated – if I
knew a person's height, I could guess their weight with more accuracy than if
I had no information. The correlation isn't perfect, however, because there is
a wide range of possible weights for any given height.

Identifying correlations between two data sets is simple, in principle. When
ψ increases, does ϕ do the same? What about when ψ decreases? The answers
to these questions can tell you about the relationship (or lack thereof) between
two data sets, but hidden in these straightforward questions are many devilish
details. Is the magnitude of the increase or decrease actually significant? What
if the data are very noisy? What if ψ and ϕ are not linearly proportional?

Remember the old saying when interpreting correlations: correlation does
not establish causation. Correlation may point to a causal link, but you'll
need additional evidence to prove a causal relationship between two quantities.
That said, many causal links have been found because a high correlation led an
analyst to thoroughly investigate the underlying processes. Furthermore, many

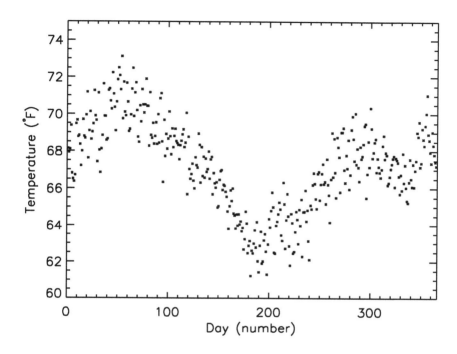

FIGURE 6.3
Mean of the average temperatures for each day of the year in the Nairobi, Kenya data set. Compare to Figure 5.14 for Addis Ababa.

supposed causal links have been discounted because no meaningful correlation could be found.

In this section, I will discuss several techniques designed to identify correlations in data sets. The techniques can be applied if ψ and ϕ are both lists or if they are both 1-dimensional arrays.

6.4.1   Scatter Plots

When analyzing jointly measured data sets, the simplest thing to do is to plot the related elements against each other. This visualization is called a scatter plot, because the points tend to be scattered around the area of the graph.[1]

The shape made by the plotted points contains information about the relationship between the two data sets. If the points fall on a diagonal line

---

[1]It is possible to convey the frequency distributions of both data sets along with the scatter plot – Tufte (2001) discusses some useful visualizations in this respect.

with a positive slope, that is evidence for a positive correlation between the two data sets. In other words, at locations when the value of one data set is relatively high, the value of the second is also likely to be relatively high. When the value of one data set is relatively low, the same is likely to be true of the other. Negative correlations between the data sets connect relatively high values of one data set to relatively low values of the other (this is sometimes known as anti-correlation). If the points fall on a straight line with a zero slope, that is indicative of no correlation between the data sets.

The tighter the points cluster to a line, the more significant the correlation between the two data sets. For this reason, scatter plots of two data sets that lack a correlation resemble circular or square clusters of points.

The scatter plots of correlated data sets do not necessarily fall on straight lines – there is a huge variety of possible relationships. For example, imagine I had a coin collection and I measured each coin's radius and surface area. If I graphed these measurements on a scatter plot, the points would fall along the curve given by Area=$2\pi\times$Radius^2, with some measurement error added on. These data sets have a positive correlation that follows a parabolic curve. Clusters of points that follow one relation in part of the plot and a second relation in another part are also possible.

The scatter plot for the heights and weights of the occupants of the classroom is shown in Figure 6.4. Notice how the three points corresponding to the adults in the room are distinct from the rest of the points in the data set.[2] I have overplotted a grey line corresponding to a 2 to 1 weight to height relationship. The points are clustered fairly close to that line, as expected from the data set's method of creation.

For another example of a scatter plot, let's plot the Nairobi temperature data set against the Addis Ababa data set. In order to construct this plot, I needed to build jointly measured data structures because both data sets have gaps, and the data is missing for the two cities on different days. There are only 1832 days with properly recorded temperatures in both cities. The scatter plot for the two temperature data sets is shown in Figure 6.5. The units of both axes are temperature, and each point on the plot represents the daily average temperature for one day in the two cities. The amorphous blob of points conveys no obvious correlation between the two temperatures (a circular distribution corresponds to no correlation). Looking more closely, there is a slight hint of an oval-like shape oriented along the lower-left to upper-right axis of the plot, but the evidence is not terribly compelling.

As briefly discussed in Section 6.2, the construction of the jointly measured data sets has the potential to introduce systematic error because there is no guarantee that the data are missing on randomly drawn days. For example, imagine if the temperature-measuring apparatus failed to record data at Addis Ababa on exceptionally cold days. Then, no points corresponding to those exceptionally cold days would appear on the plot, and important effects could

---

[2]Granted, this is not real data, but it does demonstrate the point well.

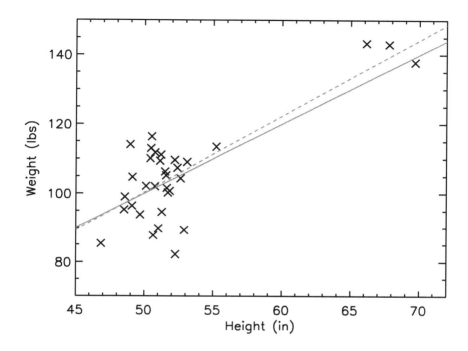

FIGURE 6.4

Scatter plot for the heights and weights of the 33 occupants of a classroom. The solid grey line depicts a 2:1 relationship between each person's weight and height. The dashed grey line depicts the best-fit linear relationship described in Section 6.6.

be missed. As usual, use your judgment as to whether the missing data is skewing your results.

For a more straightforward scatter plot to analyze, consider the elevation data sets. These two data sets already constitute jointly measured data structures, since neither data set has any gaps. I've already remarked that I expect the two data sets to be similar; let's see how that plays out on a scatter plot (Figure 6.6). Most of the points cluster around a line representing a one-to-one relationship, with some variability. The tight positive correlation between the data sets, as evidenced by the 45° line plotted over the points, is a strong contrast to the nearly circular shape of the scatter plot for the temperature data sets. The tracks of points deviating from the one-to-one correlation line correspond to physical features in one elevation data set that are different than the other data set. For example, the 5" north data set may have an elevated ridge that is not present in the equatorial data set. All in all, the evidence for

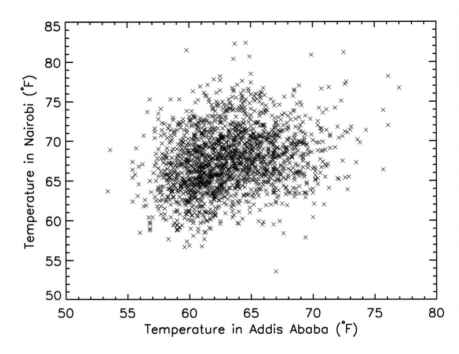

FIGURE 6.5
Scatter plot of the two temperature data sets, at Addis Ababa and at Nairobi.
Each 'x' represents the daily average temperature measured at both locations
on a single day.

correlation is much stronger in the elevation data sets than in the temperature
data sets.

6.4.2 Product-Moment Correlation Coefficients

There are many statistics used to describe the degree of correlation between two data sets. One of the most commonly used is the
<u>Pearson correlation coefficient</u>, given by:

$$\rho = \frac{\sum_{i=0}^{m-1} \left[(\phi_{z,i} - \mu_{\phi_z})(\psi_{z,i} - \mu_{\psi_z}) \right] / (m-1)}{\sigma_{m-1,\phi} \sigma_{m-1,\psi}}. \tag{6.3}$$

The Pearson correlation coefficient measures <u>product-moment correlation</u>; as
evidenced by Equation 6.3, in this type of correlation statistic each element
is compared to the mean of its data set, then the residuals are multiplied
together. Essentially, positive correlation is credited when both remainders

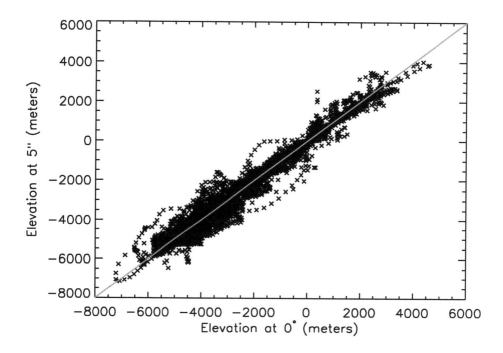

FIGURE 6.6
Scatter plot of the two elevation data sets, at the equator and 5" north of it.
A one-to-one positive correlation is marked by a solid grey line.

are above their respective means or below their respective means together.
When one is above its mean and the other is below, negative correlation is
credited.

The Pearson correlation coefficient is normalized to vary between -1 and
1, with 1 representing perfect correlation (a positive relationship), and -1
representing perfect anti-correlation (a negative relationship). If ρ is 0, then
no correlation has been revealed by this statistic. Since actual data rarely falls
exactly on any of these three cardinal values, there are some heuristic rules for
what level of correlation is "significant," but no methodology that is solidly
grounded (Press et al. 2007).

The Pearson correlation coefficient is calculated from the jointly measured
data sets $\psi_{z,i}$ and $\phi_{z,i}$, which raises an interesting question: should the means
in the numerator and the standard deviations in the denominator be calculated
from the jointly measured arrays or the original data? For data sets where no
manipulation is required to calculate the jointly measured data sets or those
constructed just by adding or subtracting a constant offset to the location
array, there will not be any difference between the statistics. For situations in

which the jointly measured data sets are subsets of the original data, there could be significant differences in the means. My recommendation would be to calculate ρ using the mean and standard deviation of the subset because that preserves the normalization of ρ. However, if the mean and standard deviation are substantially different for the jointly measured data set than for the full data set, that may be a sign that the jointly measured data set is not a random subset of the full data. This is a dangerous situation, and implies that the calculated ρ is likely not a good representation of the correlation between the full data sets.

In the simplest cases, there is a straightforward relationship between the shape of the points on a scatter plot and the value of the Pearson correlation coefficient; in fact, you can think about ρ as an attempt to reduce a scatter plot to a single number. If the points on the scatter plot are tightly clustered to a line with positive slope, that usually implies a positive ρ. Tightly clustered points near to a negative sloped line imply a negative ρ. A lot of scatter coupled with a zero-sloped line implies a ρ near zero.

Of course, there are a huge number of shapes that a scatter plot can take beyond simple linear relations. Most of these shapes don't have intuitive connections to the Pearson correlation coefficient, but that won't stop many analysts from blindly calculating ρ regardless. This is one of the repercussions of reducing a scatter plot to a single number – a lot of detail is lost! For example, there are many scatter plot shapes that can result in $\rho = 0$, but still have very interesting (and significant) relationships between the variables. Test this by calculating ρ for a 'V' shaped scatter plot. To mitigate the consequences from this loss of detail, it's never a good idea to blindly apply a correlation coefficient without looking at the data themselves. Always examine the scatter plot to confirm that ρ means what you think it means.

Calculating the Pearson correlation coefficient for the height and weight data sets (the scatter plot shown in Figure 6.4) results in $\rho = 0.8$. Intuitively, a positive ρ isn't surprising, and the scatter plot confirms it because of the positive relationship between height and weight. Certainly, all of the lower weight people have relatively short heights, and the taller adults have heavier weights. The reason ρ isn't closer to $+1$ is that some of the shorter children have relatively heavier weights.

For the elevation data sets, the Pearson correlation coefficient works out to be 0.995. This doesn't do much other than confirm the obvious intuitive result; the elevations at the equator and at 5"N have a strong positive correlation. Given the tight relationship in Figure 6.6, this value of ρ is not a surprise.

For the temperature data sets, $\rho = 0.258$. Comparing the scatter plot for these data sets (Figure 6.5) to that of the elevation data sets (Figure 6.6), the lower Pearson correlation coefficient for the temperature data sets also makes sense. Clearly, the points in the temperature scatter plot are less tightly clustered to a line than the points on the elevation scatter plot.

Note that comparing the Pearson correlation coefficients for the elevation and temperature data sets is possible because ρ is unitless. As a side note, this also means that if I changed the units of the numbers used to calculate it (such as from meters to kilometers in the elevation data sets) then ρ would not change.

6.4.3 Rank Correlation Coefficients

If you're interested in looking at some other ways to measure correlation, there are a few common statistics for <u>rank correlation</u>, a statistic that measures whether two variables increase or decrease at the same time. The technique I will discuss in depth, the <u>Kendall τ rank correlation coefficient</u> (Kendall 1938), does not depend on the magnitude of the increase or decrease, but some rank correlation methods do have this dependence.

Kendall τ is constructed elegantly. For the m pairs of elements that exist in the jointly measured data structures, I can construct $\frac{1}{2}m(m-1)$ different pairs of pairs (this is simple combinatorics). The algorithm for Kendall τ calculates the number of these pairs that demonstrate positive correlation, and the number that exhibit negative correlation. Positively correlated pairs are ones in which both ψ and ϕ increase or both decrease. Pairs in which ψ increases and ϕ decreases, or vice versa, are negatively correlated (Figure 6.7). Mathematically:

$$a \ = \ \text{\# of positively correlated pairs of pairs}$$
$$b \ = \ \text{\# of negatively correlated pairs of pairs}$$
$$\tau \ = \ \frac{a-b}{\frac{1}{2}m(m-1)} \tag{6.4}$$

Pairs of pairs where the values of ψ or ϕ do not change are "ties" and do not enter into the numerator, at least in this version of τ.[3] Notice how the magnitude of the change in ψ and ϕ does not enter into Kendall τ at all. Like ρ, $|\tau| \leq 1$, because the denominator of the fraction is just the total number of pairs of pairs.

The statistical significance of a measurement of τ can be found from the expected standard deviation of two lists that are uncorrelated. Luckily, an approximation for this quantity is available for $m > 10$ (Press et al. 2007):

$$\sigma_\tau = \sqrt{\frac{4m+10}{9m(m-1)}}. \tag{6.5}$$

Notice how this equation depends solely on the number of elements in the jointly measured arrays. Notice also how as m gets larger, σ_τ approaches 0; this makes sense, since a given value of τ is more likely to be significant if it is drawn from a larger number of pairs.

---

[3] A version of Kendall τ which does adjust the coefficient for ties is described in Press et al. (2007); in fact, several modified versions of τ are common. So, as usual, state clearly which variant you are using in your analysis, and read carefully when interpreting someone else's results.

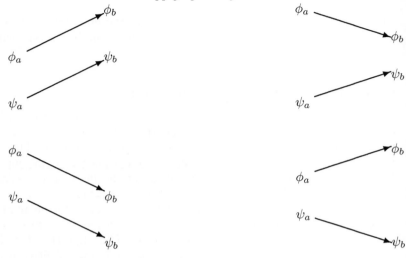

Positive Correlation   Negative Correlation

FIGURE 6.7
The two cases of positively correlated pairs of elements and the two cases of negatively correlated pairs of elements in a Kendall τ rank correlation coefficient calculation.

For the height and weight lists, Kendall τ is 0.22. The relatively low τ here is due to the number of pairs of children where the mean height-weight relationship is drowned out by the random noise added on top. Still, for $m = 33$, $\sigma_\tau = 0.12$. This means that the finding of a correlation is approximately a 2σ result.

When I calculate Kendall τ for the elevation data sets, I find $\tau = 0.9999$. As with the calculation of ρ for the elevation data sets ($\rho = 0.995$), finding $\tau \approx 1$ is wholly unsurprising. Both of these correlation statistics simply confirm the tight relationship observed in the scatter plot. Our intuition tells us that this is a statistically significant result, but calculating $\sigma_\tau = 0.004$ confirms it.

Similarly, for the temperature data sets, $\tau = 0.179$. Both ρ (0.258) and τ indicate a weak positive correlation between the temperatures in the two cities. The calculation of $\sigma_\tau = 0.016$ points out that a real correlation between the temperatures is likely. Of course, simply showing the scatter plot is much more descriptive of the actual relationship between the variables than either of the two statistics alone.

If you're interested in a rank correlation coefficient that does depend on the magnitude of the change in the two variables, there are a number of techniques available, of which the most commonly used is Spearman's ρ (Spearman 1904; Press et al. 2007).

6.5   Ratios

The <u>ratio</u> of two elements is calculated by dividing one by the other. Ratios are one of my favorite tools for revealing the relationship between two data sets. For example, consider two data sets, ψ and ϕ, functionally related by:

$$\psi = c\phi + f(t), \tag{6.6}$$

where $f(t)$ represents random noise added to each element. Imagine the analyst didn't know this underlying relationship, only having access to the two data sets. Even when the relationship between the two data sets is as simple as Equation 6.6, random fluctuations can hide those relationships in a scatter plot. However, if I calculate the ratio $r = \psi_{z,i}/\phi_{z,i}$, the points will cluster around the value c with scatter given by $f(t)/\phi_{z,i}$, making the relationship more apparent.

Ratios can be visualized in many different ways. When analyzing two 1-dimensional arrays, just plotting the ratio as a function of element number can help to reveal trends that vary as a function of position in the array. Alternatively, it can be useful to construct histograms of ratio values, which can make apparent mean and distribution properties that were otherwise hidden. Many other visualizations are possible.

It's important to realize that ratios don't have any more information than scatter plots; in fact, they have less because they reduce two pieces of information to one. However, sometimes looking at the problem in a different way (and yes, even throwing away some information) can reveal relationships that aren't apparent at first glance. The human mind works in strange ways!

Consider the weight to height ratio constructed from the height and weight data sets. Figure 6.8 shows a histogram of the ratio counts, which is clearly centered around 2.0. This is exactly as expected, since I constructed the weights by multiplying the heights by two and adding a random number. In fact, Equation 6.6 is the general form of the relationship between height and weight I applied to generate the data. Compare Figure 6.8 to the scatter plot in Figure 6.4. Note how the 2:1 weight to height trend is more difficult to see in the scatter plot due to the choice of axes, the noise, and the presence of the outliers. Although Figure 6.8 has less information, the relationship between the two variables is easily apparent to an analyst.

Figure 6.9 plots the ratio between the daily average temperature of Addis Ababa and Nairobi for the jointly measured data sets by day number. The characteristics of these data sets lead to the plot having a few strange features. For example, the large gaps where there are no points plotted are days when either the Addis Ababa or Nairobi data was missing. Overall, it is apparent that Nairobi is somewhat cooler than Addis Ababa on most days; of course, simply examining the relationship between the data set averages revealed that fact, but looking at the daily fluctuation in the ratio contains information about the variability in that relationship.

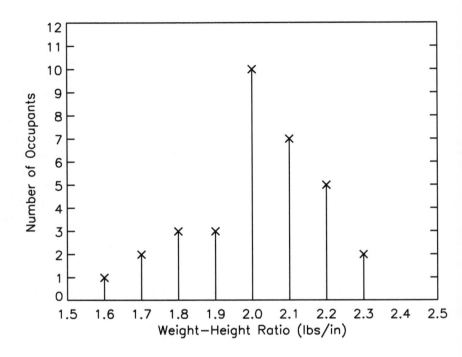

FIGURE 6.8
Histogram of the weight to height ratio of the classroom data set.

Taking the ratio of two elements measured with the Fahrenheit temperature scale is not strictly allowed, according to the rules I discussed in Section
2.2. Since 0°F does not represent the absence of temperature, it is not correct
to state that one city is twice as hot as another because there is a 2:1 ratio
between their temperatures. Similarly, it should not be argued that Addis
Ababa is 10% cooler than Nairobi as evidenced by the ratios in Figure 6.9 being centered near 0.9 – instead, the figure illustrates simply that Addis Ababa
is typically cooler than Nairobi. Contrast this finding with some of the other
data sets we've discussed, such as height and weight, where measurements of
0 lbs or 0 in. actually do represent the complete absence of those quantities.
This is the difference between interval scales, such as Fahrenheit temperature,
and ratio scales, such as height and weight.
    Figure 6.10 depicts the ratios of the elevation at the equator to the ele-
vation at 5" North latitude. Unsurprisingly, most of the ratios are clustered
around 1.0, but there are clearly a few outliers. Some of these outliers result
from ratios taken when one of the elements is small. For example, imagine if
one of the measurements was 1.0 m, and the other was 0.5 m. The absolute
difference between the two measurements is 0.5 m, which is a fairly small dif-

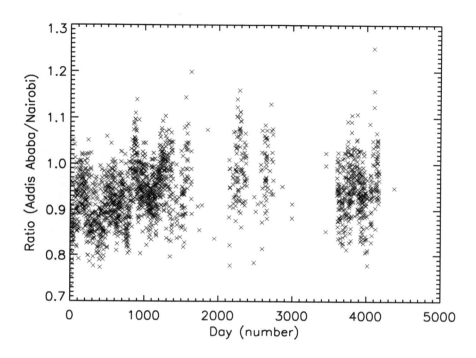

FIGURE 6.9
Plot of the ratio between the daily temperatures of Addis Ababa and Nairobi.

ference, but the ratio between the two quantities is 2:1. It is possible to filter out this effect by not plotting points where either of the measurements is 10 m or less, but if you're interested in the behavior near sea level that filter could mask the signal that you're interested in. Since I am more interested in determining whether the majority of elements cluster around a 1:1 ratio, I chose the *y*-axis range of Figure 6.10 to cut off the very large and small ratios and focus on elements near 1:1.

There are occasions when examining the ratio between two variables will simply reinforce the trends observed in a scatter plot, but will not offer any new insight. That seems to be the case with the elevation data set; personally, I find the scatter plot in Figure 6.6 more illuminating than the ratio plot of Figure 6.10. In fact, I find the ratio plot for the elevation data set so noisy and confusing that I wouldn't use it as part of a presentation.

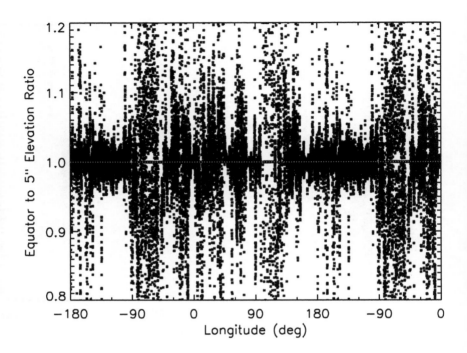

FIGURE 6.10
Plot of the ratio between the elevation at the equator to the elevation at 5"
North latitude.

6.6   Fitting Related Lists and 1-Dimensional Data Sets

In Section 5.6, I discussed fitting a functional form to a 1-dimensional data
set. Here, I will discuss how fitting is used to model the relationship between
two related data sets, $\psi_{z,i}$ and $\phi_{z,i}$, similarly to how in the earlier discussion I explored the relationship between the independent variable x and the
dependent variable ψ.

Essentially, the trick in fitting related data sets is to choose one to be
the independent variable and the other to be the dependent variable. Using
the height and weight data sets as an example, I can choose the height to
be the independent variable and the weight to be the dependent variable.
Then, I can fit a linear functional form to the data using the S^2 least squares
formulae described in Section 5.6.1 for a straight line. Essentially, I'm fitting
the following functional form:

$$\text{Weight} = m \times \text{Height} + b. \tag{6.7}$$

This fit results in $m = 2.2 \pm 0.3$ lbs/in. and $b = -9 \pm 16$ lbs, where the quoted errors are the 1σ uncertainties defined in Equations 5.8 and 5.9. Note how the values of the linear relationship I used to create the weight data set, $m = 2$ and $b = 0$, are within the 1σ error bars found by the fit. The best-fit linear relationship is plotted on top of the scatter plot in Figure 6.4.

Note how this fit has a range of validity given by the range of the elements in height. It wouldn't make any sense to use the fit to estimate the weight of a person that was 30 in. tall because there just isn't any data there constraining the fit. A similar warning applies to estimating the weight of someone 80 in. tall. The fit can also be used to estimate someone's height if given their weight – the range of validity can also be expressed in pounds.

6.7 Correlation Functions

So far, none of the statistics that I've discussed have utilized the order of the elements. In other words, I could scramble the order of ψ, and as long as I scrambled ϕ in the same way, the correlation coefficients and scatter plots would look exactly the same. This implies that, for 1-dimensional data sets, I haven't taken advantage of all of the information I possess. The techniques I will describe in this section will take advantage of the information contained in the measurement location arrays, so they are appropriate for use on 1-dimensional data sets but not lists.

Frequently, analysts need to investigate is whether or not two 1-dimensional data sets have similar shapes. One of the most commonly used methodologies for this type of analysis is the correlation function, drawn from the field of digital signal processing (Oppenheim et al. 1999). The correlation function can be used to identify similarities between two data sets or to look for repetition at different locations in a single data set. In the first case, the algorithm is sometimes called the cross-correlation function and the second case is generally referred to as the autocorrelation function.

The purpose of a cross-correlation function is to compare overlapped copies of two 1-dimensional data sets that are offset by some amount of the measurement location array. The reason it's a "function" is that the degree of similarity is expressed as a function of that offset. To start, let's limit this discussion to the simplest case: jointly measured 1-dimensional data sets with m elements, where the spacing between the elements is consistent. In that case, the algorithm for the cross-correlation function, denoted by X, is given by:

$$X_j = \sum_{i=0}^{m-1} \phi_{z,(i+j)} \psi_{z,i} \qquad (6.8)$$

for $0 \leq j \leq m - 1$. Essentially, X_j has its largest magnitudes when the elements of ψ with large magnitude are multiplied by the elements of ϕ with large magnitude. When those same large magnitude parts of ψ are lined up with the small magnitude parts of ϕ, then X_j will be of smaller magnitude. Note how strange the units are in this version of the cross-correlation function, because the two 1-dimensional data sets are not required to have identical units. That means the units of X_j are simply the units of ϕ and ψ multiplied together. Thus, the magnitude of X_j has meaning only with respect to other values of X_j in the same correlation function; that is a less than ideal quality.

The sign of the correlation function, as defined in Equation 6.8, requires some interpretation from the analyst. Most importantly, elements of X_j that are negative are not necessarily "negatively correlated." Imagine the cross-correlation between the functions $\psi = \sin(x) + 1$ and $\phi = \sin(x) - 1$. ψ is everywhere greater than or equal to zero, and ϕ is everywhere less than or equal to zero. Therefore, X_j is everywhere less than or equal to 0. However, the functions are quite similar in shape. Choosing a different normalization for the cross-correlation function can alleviate this difficulty, as I will discuss below.

Leaving aside these complications, there is another subtlety in how the cross-correlation function is defined; what does it mean for the array $\phi_{z,i}$ to be defined for i larger than $m - 1$? In other words, imagine ϕ and ψ of the same length; when the arrays are offset and some of the elements in both arrays don't have a partner, what should those elements be multiplied by?

There are a couple of reasonable answers to that question. Could they be multiplied by 0? Well, that would bias the cross-correlation function to peak when there was no offset, and fall off with every element shifted, hiding the trends I am are trying to uncover. Instead, the data structures are usually treated as periodic (you can also think about this as assuming circularity in the data structure). Some analysts do multiply elements without a partner by 0; this works best when the offset is a small fraction of the overall length of the data sets.

As you can see from Figure 6.11, the periodicity assumption can result in some strange-looking functions due to discontinuities. Imagine a function that has a yearly cycle embedded within it; if the data recording that function is cut off midcycle, the periodic function can have a shape that jumps around in an unrealistic way. Be aware of this as you examine the results of the cross-correlation function.

As usual, there are a few different normalization conventions for the cross-correlation function. Consider the following function:

$$\hat{X}_j = \frac{\sum_{i=0}^{m-1} \left(\phi_{z,i+j} - \mu_{\phi_z} \right) \left(\psi_{z,i} - \mu_{\psi_z} \right)}{\sqrt{\sum_{i=0}^{m-1} \left(\phi_{z,i} - \mu_{\phi_z} \right)^2 \sum_{i=0}^{m-1} \left(\psi_{z,i} - \mu_{\psi_z} \right)^2}} \tag{6.9}$$

where μ_{ψ_z} and μ_{ϕ_z} are the means of the jointly measured arrays. Note how this convention turns $\hat{X}_j$ into a dimensionless quantity. In the case of perfect cross-correlation ($\psi_z = \phi_z$) and zero lag, then $\hat{X}_0 = 1$. Similarly, when

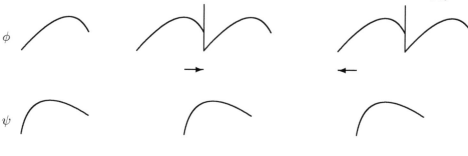

<table>
<tr><td>Zero Offset</td><td>Positive Offset</td><td>Negative Offset</td></tr>
</table>

FIGURE 6.11
Illustration of the periodicity used to calculate the cross-correlation function.
In this diagram, ϕ and ψ have been given arbitrary shapes. To calculate the
cross-correlation function, each element in ψ is multiplied by the element in ϕ
directly above it; elements of ϕ that don't have corresponding elements of ψ
located directly below don't contribute to X. The horizontal arrows show the
direction and amplitude of the shift. The vertical line marks the discontinuity
that can result when periodicity is assumed. Note how strange the function
by which ψ is multiplied can look with both positive and negative offsets.

$\psi_z = -\phi_z$ and zero lag, then $\hat{X}_0 = -1$. In this way, Equation 6.9 is normalized
to resemble the correlation statistics discussed in Section 6.4, though conceptually it is measuring a different aspect of the relationship between the data
sets. It also solves another of the problems inherent in the original definition of
the cross-correlation function by subtracting out the mean from each function.

When interpreting a cross-correlation function, try to use your eye to pick
up the reasons for its maxima and minima. Where do the functions have
similar shapes, and how far apart are those locations? Are there outliers that
are dominating the cross-correlation? This can happen when elements with
large magnitudes are multiplied by each other; this behavior can be hidden
using smoothing techniques described in Section 5.5.

For the elevation data sets, the assumption of periodicity holds without
issue, because these data structures are circular. The cross-correlation function
(as defined by Equation 6.9) for the two elevation data sets is shown in Figure
6.12. While you're looking at that plot, refresh your memory as to what the
elevation functions themselves look like with Figure 6.1. There are three major
landmasses in this plot, located around longitudes $-70°$ (South America),
$30°$ (Africa), and $100°$ (Indonesia). When these landmasses are aligned by
the proper offsets, the oceans surrounding them also align and the cross-
correlation function will peak. The exact offset where the peak occurs depends
on the interplay between the width of the oceans and the continents. However,

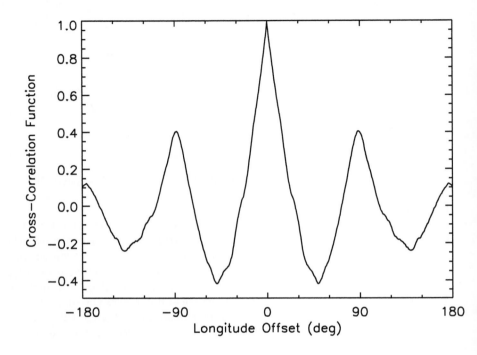

FIGURE 6.12
Plot of the cross-correlation function of the elevation at the equator to the elevation at 5" North latitude.

I can use the locations of the continents to understand the origin of each of the peaks in the cross-correlation function. The peaks near −180° and 180° stem from the overlap between South America and Indonesia. The peaks at −90° and 90° are due to the overlap between South America and Africa and the overlap between Africa and Indonesia. The 45° half width of the peak around zero offset relates to the typical width of the continents.

  The reason Figure 6.12 looks symmetric around 0° longitude is because the two elevation functions are nearly identical. In this case, the cross-correlation function isn't actually symmetric, it just appears so at the scale of the figure. In actuality, the function on either side of zero offset is slightly different, but this example has a lot in common with the autocorrelation function discussed in an upcoming inset.

  If you're dealing with a data set where calculating the cross-correlation function takes a significant amount of computational runtime, there are efficiency gains that can be realized through the connection between correlation functions and the Fourier transform. This connection allows you to replace the n^2 algorithm of Equation 6.8 with the $n \log_2 n$ algorithm of the FFT discussed in Section 5.8.1.1. More information on how to incorporate the FFT into calculating the cross-correlation function can be found in Press et al. (2007).

Autocorrelation Function

The autocorrelation function has the same structure as the cross-correlation function discussed in the previous section; the only difference is that instead of comparing two different functions, it compares a single function to itself. Thus, autocorrelation would have more appropriately belonged in Chapter 5, however I chose to discuss it here because of its direct relation to the cross-correlation function.

Following Equation 6.8, the basic form of the autocorrelation function is given by:

$$Y_j = \sum_{i=0}^{m-1} \phi_{z,i+j}\phi_{z,i}. \tag{6.10}$$

Of course, the variety of autocorrelation normalizations is the same as the variety of cross-correlation normalizations, including a version of the normalized, mean-adjusted form of Equation 6.9.

Autocorrelation functions can pick up some of the same features that result in peaks in the frequency analyses from Section 5.8. For example, imagine a simple sine wave function. A Fourier analysis would reveal a peak in the spectrum corresponding to the frequency of the sine wave. Similarly, when the autocorrelation function is calculated with an offset equal to an integral number of wavelengths, Y_j will contain a local maximum.

I recommend using the autocorrelation function to supplement a frequency analysis. It's especially convenient when looking for shapes in a single function that don't repeat regularly or have patterns that don't look like simple trigonometric functions. It also has the additional (and nontrivial) benefit that an offset is a more straightforward concept to explain to a non-technical audience.

The autocorrelation function is symmetric around zero offset, for the same reason that the cross-correlation function of the elevation data sets was nearly symmetric: there isn't a difference between a positive and negative offset when the functions you are comparing are identical.

The same calculation efficiency gains that are attainable by using the FFT with the cross-correlation function can be had with the autocorrelation function.

6.8 Exercises

1. Download the ETOPO1 elevation data set for measurements at 45°
   N. Construct jointly measured data structures with the elevation at
   the equator, then build a scatter plot. Compare and contrast this
   scatter plot with the one constructed in Figure 6.6.

2. Calculate Pearson ρ and Kendall τ for the elevation data sets at the
   equator and at 45°N. Compare and contrast these statistics with the
   quantities calculated for the elevation at the equator and at 5" N.

3. Calculate and interpret the normalized cross-correlation function
   for the elevation data sets at the equator and at 45°N.

4. Fit the relationship between the elevation data sets at 0°N and 5"N
   using a straight line. Plot that relationship on top of a scatter plot
   of the data. Determine the uncertainty in the fitted parameters.

5. Fit the relationship between the two temperature data sets using a
   straight line. Plot that relationship on top of a scatter plot of the
   data. Determine the uncertainty in the fitted parameters. For this
   fit, use Nairobi as the dependent variable and Addis Ababa as the
   independent variable.

6. Calculate the height to weight ratio of the height list and plot the
   ratios in a histogram similar to Figure 6.8.

7. Repeat the straight-line fit of the height and weight data using
   height as the dependent variable and weight as the independent
   variable. Calculate the uncertainty in the fitted parameters. How
   do these fitted parameters relate to the ones from the fit described
   in Section 6.6?

7

2-Dimensional Data Sets

7.1 Introduction

Some data sets fit much more naturally into a multidimensional structure than into related lists or 1-dimensional data sets. The benefit of organizing appropriate data into a multidimensional structure is that, by taking advantage of relationships between the elements, analysts can apply an additional suite of techniques. These techniques will sometimes reveal features in the multidimensional data set that wouldn't otherwise have been apparent. Multidimensional data sets should not be intimidating – in fact, people use several multidimensional data sets in their everyday lives. For example, many tables fall into this category of structure, such as a table of weather data in which the horizontal headings are the day of the week, the vertical headings are the time of day, and the table elements are temperature or the chance of precipitation.

In this chapter, I will discuss 2-dimensional data sets, which (obviously) are the simplest type of multidimensional data set. A 2-dimensional data set is a structure in which each element is indexed in two independent dimensions. Studying 2-dimensional data sets is useful in and of itself because much real world data is well-fit by this type of structure, but it will also bring to light concepts that aid in analyzing even higher-dimensional data sets.

The rules that govern the structure of 2-dimensional data sets are similar to those of 1-dimensional data sets, with a few subtleties related to the addition of the extra dimension of measurement locations. The elements of a 2-dimensional data set should have the same dimensionality – in other words, they should all be prices, voltages, distances, or whatever other quantity you have measured. Instead of a single array of measurement locations, 2-dimensional data sets have two, one for each dimension. Like those of 1-dimensional data sets, the measurement location arrays of a 2-dimensional data set are typically spatial or temporal. Though the dimensionality of each individual set of measurement locations should be consistent, the dimensionalities of the two don't necessarily need to match each other.

The formalism of 2-dimensional data sets is similar to that of 1-dimensional data sets. Call the elements of the data set $\psi_{i,j}$, where the i and j indices are connected to the two measurement location arrays, x_i and y_j. The indices start at 0 and run to $m-1$ and $n-1$, respectively, to allow for the possibility that the two measurement location arrays have differing numbers of elements. Therefore, the 2-dimensional data set ψ has a total of mn elements.

There are very few restrictions on the measurement location arrays x and y. As was the case for 1-dimensional data sets, for 2-dimensional data sets there is no requirement that the x and y location arrays be evenly spaced. Furthermore, the spacing in the two measurement location arrays does not need to match. However, the spacing of the measurement location arrays will influence which techniques can be used.

Some examples of 2-dimensional data sets are:

- A greyscale image, with the elements $\psi_{i,j}$ corresponding to the shading of the pixel located at x_i and y_j;

- The temperature at some specific time at a grid of locations around the Earth, with x_i corresponding to latitude and y_j corresponding to longitude;

- The temperature along the equator on a grid of longitudes given by x_i measured at a series of times y_j; and

- The number of people served by a fast-food restaurant over the past year, totaled by day of the week in x_i and hour of the day in y_j.

To some degree, the relationship between the two measurement location arrays' dimensionalities governs which techniques can be appropriately applied to a 2-dimensional data set. In cases when the dimensionalities of the two measurement location arrays match, e.g., when both are spatial in nature, I call the data set's dimensionality homogenous. In the spatial case, you can think of these data sets as 2-dimensional maps of whatever quantity the elements are measuring. For example, one measurement location array could measure distance along a north-south axis and the second along an east-west axis. When the dimensions of the two measurement location arrays don't match, e.g., when one of the dimensions is spatial and the other is time, I call the data set's dimensionality heterogenous. This common combination of measurement location arrays is known as spatiotemporal, though other heterogenous combinations are possible.

One confusing situation that can occur is when both measurement location arrays are spatial, but represent very different scales. For example, consider a soil scientist constructing a 2-dimensional data set of temperatures (the elements of the data set) measured along a straight line (the first measurement location array) at a series of depths (the second measurement location array). If the first measurement location array ranges from 0 to 1000 m and the second ranges from 0 to 5 cm, I would consider this data set's dimensionality to be heterogenous rather than homogenous, even though both measurement location arrays are spatial.

Why would that type of data set be heterogenous? For a data set's dimensionality to be homogenous, in addition to the requirement that the measurement location arrays match dimensionality, there also must exist a single distance measure, whether spatial, temporal, or some other quantity, that connects the two dimensions in a way that makes analytic sense. On a simple 2-dimensional map this distance measure clearly is reasonable – I can measure a physical distance between two elements that differ in both longitude and latitude. In a spatiotemporal data set, this distance measure does not exist – I have to say that two elements are separated by some amount of distance and also some amount of time. In the soil scientist example, I can construct a single distance measure, but it's totally useless for me to say that two elements are separated by 400.01 m – it makes much more sense for me to keep the two measurement location dimensions separate and say that the elements are separated by 400 m along the line (the first measurement location array) and 1 cm in depth (the second). Analytic judgment is required to make this categorization.

7.2 Example 2-Dimensional Data Sets

In this chapter, I will examine three 2-dimensional data sets.

7.2.1 Elevation Map

The first example data set I will examine is a 2-dimensional elevation map drawn from the ETOPO1 data set, the same source as the 1-dimensional equatorial data set (Section 5.3.2) and the related elevation data set at 5" N (Section 6.3.2) (Amante and Eakins 2008). I downloaded the portion of the map ranging from $-76°$W to $-71°$W and from $38°$N to $42°$N. This data set has a size of 301 (East-West) by 241 (North-South) elements. In this subset of the global map, the elevations range from 3513 m below sea level to 1066 m above. Geographically speaking, this data set includes part of the Atlantic Ocean and several Northeastern American states. I will call this the elevation map.

7.2.2 Greyscale Image

The second example 2-dimensional data set is a greyscale image. In a greyscale image, each element (known as a pixel) has a ratio-scale value ranging from 0 (black) to 1 (white). The particular image I will be analyzing is a geometric pattern that I created. It spans 100 pixels in the horizontal dimension and 60 pixels vertically.

7.2.3 Temperature Anomaly Global Maps

The National Climatic Data Center, part of the National Oceanic and Atmospheric Administration, has compiled and released global maps of temperature anomalies for every month since January 1880 to December 2010. This data set is known as the Global Historical Climatology Network Temperature Database (Jones and Moberg 2003; Peterson and Vose 1997). The data set is created from station measurements on an irregularly spaced 2-dimensional grid, which are then interpolated onto a regularly spaced 2-dimensional grid with some missing data. The data set for each month consist of a 72 (longitude) by 36 (latitude) grid, with missing data flagged by a large negative number. Note that the temperature data is circular in one dimension (East-West) but not in the other (North-South). I will call this the climate data set.

Each element is presented in terms of a temperature anomaly, a technical concept from climate science. A temperature anomaly is defined as the difference between the average temperature measured at a given location over some time period (e.g., the month of January 1880) and the baseline temperature at that location.[1] This definition implies that temperature anomalies are an interval scale. Temperature anomalies that are above zero correspond to average temperatures warmer than the baseline and those that are below zero to average temperatures cooler than the baseline.

This large data set is a powerful source for investigating long-term climate trends. In fact, the monthly spacing between elements in the temporal dimension is much shorter than the time period I am interested in analyzing. Therefore, to reduce the size of the data set, I did some preprocessing to the data – I grouped the monthly reports into maps of decadal averages. For example, I constructed the first decadal average by taking the data from January 1880 through December 1889 and averaging the temperature anomalies at each grid point over the 120 elements. If any of the monthly reports in that decade were missing an element, I flagged that element as missing in the average. In other words, the temperature anomaly had to be measured every month through the decade to be included in the average map for that decade.

After this preprocessing is completed, the climate data set is a 3-dimensional data set with two spatial and one temporal dimension. In this book, I will limit the analysis of this data set to techniques that compare related 2-dimensional structures, which I will create by extracting data from particular decadal averages. More sophisticated techniques that incorporate all three dimensions are possible.

---

[1] For this data set, the baseline temperature is defined as the average temperature measured at that location between 1960 and 1990.

7.3 Restructuring Data Sets

Astute readers may have noticed that some of the example 2-dimensional data sets in the introduction and in the previous section can be restructured into data sets with other structures. Indeed, analysts have freedom to restructure data into convenient forms, and taking advantage of this freedom sometimes allows you to apply techniques that would not otherwise be possible.

As examples of restructuring, the fast-food data set described in Section 7.1 could have originated in either of these two forms:

• Two related lists, structured similarly to the height and weight lists from Section 6.3.1, wherein each element corresponds to a single customer, with the first list containing the day that the customer was served and the second containing the hour, or

• A 1-dimensional structure in which the measurement location array contains the day of the week and the hour, and the elements in the list contain the total number of customers served.

Similarly, the greyscale image could have been organized into three related lists, where the first contained the pixel's ratio-scale shade, the second the pixel's x location, and the third the pixel's y location. Or, it could have been a weirdly structured 1-dimensional array, with the pixel shadings contained in ψ and the measurement location array corresponding to pixel number, starting from the upper left, increasing rightward, and moving down to the next line when the rightmost point in a line is reached.

The bottom line is that the original form in which an analyst encounters data usually has more to do with the way the data were collected than with what makes analytic sense. For example, consider two methods by which the fast-food data could have been collected. First, the manager could have constructed a table with axes corresponding to day of the week and hour of day, then asked the workers at the register to draw a hash mark on the table for each sale. After totaling the hash marks, the manager would have a 2-dimensional data set. Alternatively, the manager could have asked the workers at the register to write down in list form the day and hour of each transaction – this would result in two related lists. These two methods record the same information, but do so with very different structures.

Analysts should restructure data sets so the analytic techniques that will answer their questions of interest can be applied. It is possible that different data structures for the same data will be necessary to apply different techniques. For example, if I wanted to find frequency patterns in the fast-food data I would probably use the 1-dimensional structure, but the 2-dimensional structure makes it easier to see which days and times are busiest.

As another example of restructuring, consider the elevation map. I originally downloaded the data as three related lists: the first list contained the longitudes, the second contained the latitudes, and the final list contained the elevation at that latitude-longitude location. Each list had 72,451 elements. I restructured these lists into a 2-dimensional structure, with the x measurement location array containing the longitudes and the y measurement location array the latitudes. I did this because this structure is more conducive to the techniques I am interested in applying.

Restructuring can also change a 2-dimensional data set into a set of related lists. In Chapter 3, I discussed a variety of techniques that could be applied to lists, such as the mean, standard deviation, and histograms. Restructuring allows an analyst to apply these same techniques to a 2-dimensional data set.[2] Keep in mind that if the data vary strongly by region, these techniques can sometimes be difficult to interpret – for example, the median of the elevation map would be substantially different if calculated solely for the southeast quadrant, which is entirely ocean. Nevertheless, these list techniques are a good starting point for analyzing a 2-dimensional data set.

Similarly, 1-dimensional techniques that I discussed in Chapters 5 and 6 can be applied to individual rows or columns of a 2-dimensional data set. In fact, this is exactly what I did in the previous chapter when I compared the elevation data sets, which are rows in a much larger elevation map. Using these techniques can produce helpful results, but the techniques often fail to take full advantage of the relationships between the elements.

7.4   Analogs of 1-Dimensional Techniques

Many of the 1-dimensional analytic techniques I discussed in Chapter 5 have analogs that can be applied to higher dimensional structures. In this section, I will discuss a few techniques that can be used on a 2-dimensional data set and take full advantage of that data set's structure.

7.4.1   Interpolation in Two Dimensions

As was the case for 1-dimensional data structures, sometimes you need an estimate of ψ at a location that falls in between the measurement location array coordinates with known values, (x_i, y_j). For both homogenous and heterogenous data sets, interpolation can be used to provide this estimate.

Let's examine the simplest case of a 2-dimensional data set with evenly spaced measurement location arrays. Instead of falling between two elements with known values as in the 1-dimensional case, a point can fall inside or on the boundaries of a rectangle formed by four elements with known values.

---

[2]When calculating these statistics for a multidimensional data set using a computer, the structure itself is usually not transformed. The elements are just cycled through as if they belonged to a list.

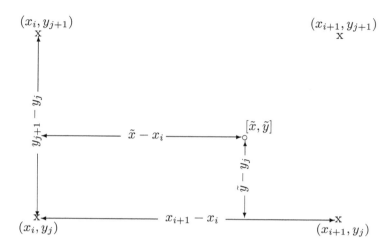

FIGURE 7.1
A schematic of the point locations and distances involved in bilinear interpolation. The four corner elements with known values of ψ are marked with x's. Measurement location array coordinates are enclosed in parentheses, and distances are written with arrows emerging leftward and rightward. The point of interest is marked with a small circle and its coordinates are in square brackets.

In general notation, these elements are $\psi_{i,j}$, $\psi_{i+1,j}$, $\psi_{i,j+1}$, and $\psi_{i+1,j+1}$ and are located at the positions (x_i, y_i), (x_{i+1}, y_j), (x_i, y_{j+1}), and (x_{i+1}, y_{j+1}), respectively. The point I am interested in interpolating the value of, $\tilde{\psi}$, is located at the position $(\tilde{x}, \tilde{y})$.

Just as was the case with 1-dimensional interpolation, a 2-dimensional interpolation can be calculated using a variety of algorithmic techniques. The 2-dimensional analog of the simple 1-dimensional linear interpolation we discussed in Section 5.4 is known as bilinear interpolation.

Bilinear interpolation is accomplished in two steps. First, calculate the fractional position in each dimension of $\tilde{\psi}$ between the measurement location array positions with known values of ψ:

$$
\begin{aligned}
f_x &= (\tilde{x} - x_i)/(x_{i+1} - x_i) \\
f_y &= (\tilde{y} - y_j)/(y_{j+1} - y_j)
\end{aligned}
\tag{7.1}
$$

These fractions f_x and f_y are dimensionless numbers between 0 and 1. The points and lengths involved in bilinear interpolation are illustrated schematically in Figure 7.1.

The second and final step of bilinear interpolation is to calculate the estimate $\tilde{\psi}$ using a weighted average of the values of the points at the four corners of the rectangle. The components of the average are weighted by the proximity of the desired point to each of the corners:

$$
\begin{aligned}
\tilde{\psi} \ = \ & (1 - f_x)(1 - f_y)\psi_{i,j} + f_x(1 - f_y)\psi_{i+1,j} \\
& + (1 - f_x)f_y\psi_{i,j+1} + f_xf_y\psi_{i+1,j+1}.
\end{aligned} \tag{7.2}
$$

In other words, the closer the point is to one of the corners, the stronger the weight the value of ψ at that corner gets.

Because it only takes three points to define a plane, bilinear interpolation is not guaranteed to connect all four corners of the rectangle of elements with a flat surface. In fact, the surface that connects the four elements is actually curved when the four elements don't fall on the same plane, so don't assume that bilinear interpolation means "planar interpolation." Check the limits of bilinear interpolation to make sure that when the point $(\tilde{x}, \tilde{y})$ falls directly on top of one of the corners, the value of ψ at that corner is returned. Additionally, when the point $(\tilde{x}, \tilde{y})$ is on one of the boundaries of the rectangle, Equation 7.2 reduces to the 1-dimensional linear interpolation between those two elements.

Bilinear interpolation isn't the only algorithm for 2-dimensional interpolation – splines and higher order polynomials can also be used to provide estimates of $\tilde{\psi}$. As for the 1-dimensional case, these functional forms will give different estimates for $\tilde{\psi}$, and it isn't always clear which estimate is closest to the real value at the point of interest. More complicated algorithmic techniques can provide some convenient properties, such as smooth derivatives when moving from one rectangle of known elements to another (the 2-dimensional analog of 1-dimensional spline interpolation). These properties often are accompanied by additional computational complexity. Additional interpolation techniques, along with interpolation for irregularly spaced multidimensional data structures, are discussed in (Press et al. 2007).

The same caveats that apply to 1-dimensional interpolation also apply to the 2-dimensional version. For example, the interpolation has to make sense given the data that you are working with – don't interpolate between two daily maximum temperatures and expect to accurately estimate the temperature in the middle of the night.

7.4.2 Contour Plots

It is very difficult to pick out the important features in a 2-dimensional data set by looking directly at the raw data in numerical form, such as would be displayed in a spreadsheet. Contour plots aid in data interpretation by helping analysts to visualize 2-dimensional data sets. A good contour plot will naturally draw attention to the peaks and valleys in the data, as well as any other interesting features. They are commonly used with ratio and interval scale elements, but can also be used with interval and categorical scales. Contour plots can be constructed from 2-dimensional data sets with regularly and irregularly spaced location arrays. Everyday uses of contour plots abound, including weather maps, topographical hiking maps, and population density maps.[3]

Essentially, a contour plot is the 2-dimensional equivalent of the simple display of a 1-dimensional data set with location on one axis and element value on the other. In contrast, the axes of a contour plot are the location arrays x and y. The coordinate (x, y) on the contour plot is encoded with the value of ψ at that point using a combination of graphical techniques. Different levels of ψ are typically denoted by lines and color shading. There is considerable freedom for the analyst to choose a set of contour levels that emphasize the behavior in ψ that he or she would most like the reader to notice.

For complex contour plots, it can be difficult for the reader to understand the relationship between the value of ψ and the contour levels. Two simple techniques are available to make this easier: color bars and contour labels. Usually used with contour plots in which the various levels are primarily marked by color or shade, a color bar is a structure offset from the main plot which depicts the relationship between the color of a level and its value. Contour labels, which can be used in conjunction with a color bar or alone, are labels where the values of some or all of the contour levels are written directly on the contours themselves. Contour labels run the risk of making a contour plot too busy and cluttered, so a lot of care is needed to ensure the labels are placed intelligently.

These days, contour plots are almost always crafted with computer algorithms rather than drawn by hand. Drawing a contour map on top of a grid of points is actually a complicated interpolation problem for which there is no single correct answer – interpolation is required because it is used to estimate where between a set of grid points a contour line is drawn. Software packages apply a variety of algorithms to draw contours so there is no guarantee that contour plots drawn with different packages will produce identical results.

Figure 7.2 depicts a contour plot for the elevation map. The horizontal and vertical axes are labeled with degrees of longitude and latitude, and the axes have been scaled so a degree in longitude is the same length as a degree in latitude. This is called a equirectangular projection; it does not accurately portray distances because degrees of longitude and latitude do not correspond to the same physical length (except at the equator). By using filled contours above sea level and line contours below, I've utilized a contour style that strongly differentiates between locations above and below sea level. I've chosen the contour levels such that contours above sea level are more closely spaced (every 200 m) than those below sea level (every 500 m) both because of the larger range of values below sea level ($\approx$3000 m vs. $\approx$1000 m) and because I've judged the fine details of features above sea level to be more interesting than those below sea level.

---

[3]An example of a contour map on an irregular grid is a national temperature map. These maps are created from a grid of weather stations, typically found in cities, which are irregularly distributed around the country.

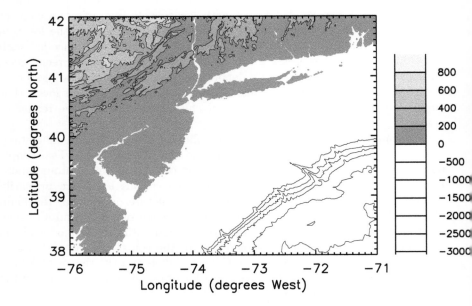

FIGURE 7.2
Contour plot of the 2-dimensional elevation map. Below sea level, contours are drawn with solid lines every 500 m. Above sea level, filled contours are drawn every 200 m. The darkest filled contours are at 0 m, with increasing elevation marked by lighter shades of grey. Contours above sea level are accentuated with a black line. The color bar on the right shows the relationship between the elevation (in meters) and the contour levels.

The New York, New Jersey, Connecticut, and Rhode Island coastlines, the increase in ocean depth in the southeast, and the mountains in the northwest are all clearly visible on the contour plot, but some of the smaller scale geographic features (such as Fire Island, a long and thin landmass along the southern shore of Long Island, which itself stretches from −72° to −74° W) do not appear at all. This behavior occurs because the spacing of the data set is too sparse to properly map relatively small or thin areas. This is a lesson in interpolation – when there are serious consequences to the precise value of an interpolation, such as whether a given point is above or below sea level, features with a scale smaller than the data's spacing may be missed entirely if their appearance depends on the precise value of the element, such as whether a given point is above or below sea level.

Note how one consequence of my choice of contour markings is that it is not possible to determine whether the small closed contour at $-72°$W and $38.5°$N corresponds to an increase or decrease in depth. From looking directly at the data, it turns out that this is a small depression in the sea floor. By leaving this ambiguity in the contour's value, I made a mistake in drawing the contour plot. I did not correct the error because I wanted to demonstrate how careful you need to be when making these plots – there is a lot of detail to pay attention to. In fact, similar concerns could be raised about many of the contours drawn below sea level.

In Section 7.5.1, I will discuss the climate data set and display more examples of contour plots. The climate data set is a particularly good example of how to visualize a structure with missing data.

7.4.3   Smoothing in Two Dimensions

In Section 5.5 I discussed the application of smoothing techniques to 1-dimensional data, including the running mean and median smoothing. These techniques can be used to a similar effect (with some modifications) on data sets with multiple dimensions.

When using these techniques on data sets with multiple dimensions, the process of calculating the mean or the median of a group of numbers is the same as it was for lists and 1-dimensional data sets. However, there is some subtlety pertaining to the shape and size of the group of elements that the functions are performed over. I discussed this idea for 1-dimensional data sets, using the concept of the half width, w_h to characterize the window over which the smoothing occurs. For homogenous 2-dimensional data sets, it is appropriate to change the half width into a radius ("half diameter") and include all of the elements that fall within a circular window. This generalizes into a spherical window for 3-dimensional homogenous data sets, and so on.

For heterogeneous 2-dimensional data sets, this simple generalization does not work because there is no distance measure connecting the two dimensions. This issue can be avoided entirely by smoothing over only one of the dimensions, but in practice many analysts use 2-dimensional windows with heterogenous data. The scales in each of the two dimensions can be chosen independently, such that the half width of the window becomes a 2-dimensional vector. There isn't a hard and fast rule for the shape of the window in this case, so the analyst must make a decision. Plus signs, rectangles, and diamonds are just a few of the possibilities, but there is an unavoidable arbitrary aspect to the choice. I suggest making sure that the results are robust to a choice of two or three different windows over which to perform the smoothing function.

Just as was the case with 1-dimensional data sets, it is possible to construct more methodologically complicated smoothing functions. One such function calculates the running mean as a weighted average, usually with the smaller weights located nearer to the edges of the window. Downweighting points towards the edges of the window can help reduce the impact of outliers on the running mean, since they enter the calculation with less influence. The outliers will still dominate when they are near the center of the smoothing shape, however.

To illustrate how smoothing works with a 2-dimensional data set, I've applied median smoothing to the elevation map. In this case, choosing the scale for smoothing has some intricacies, because the physical distance corresponding to some number of degrees of longitude is shorter than the physical distance over the same number of degrees of latitude. Each minute of latitude corresponds to 1.85 km, and each minute of longitude corresponds to 1.46 km at 38°N or 1.38 km at 42°N. Therefore, if I want to smooth using a physical distance as a half width, I must determine the pattern of elements within that radius. So if I smooth on a 3 km scale, that includes the central element plus 1 element in all 8 compass directions, as well as additional elements due east and west for a total of 11 points.[4] I avoided edge effects with this filter by using a data set slightly larger than the latitude and longitude boundaries of my original map, but many data sets require dealing with edge effects similarly to the 1-dimensional case.

So what has all this effort produced? As shown in Figure 7.3, applying the median filter to the elevation map has removed some of the fine-grained detail that was present in the original version of the data. For example, the valleys corresponding to the Hudson and Connecticut Rivers (stretching north-south near −74° and −72° W, respectively) have been made much more shallow and the Delaware River (located along a northeast-southwest path near 40° N) is noticeable for a shorter path. The larger scale features are still quite recognizable. In essence, I've de-emphasized the small-scale features to draw more attention to the large-scale trends.

7.4.4   Frequency Analysis in Two Dimensions

The process of using frequency analysis to identify repeating oscillations in 2-dimensional data sets has many similarities to its use with 1-dimensional data sets, as covered in Section 5.8.

One method of detecting oscillations in a 2-dimensional data set is to perform a series of 1-dimensional frequency analyses along one of the dimensions. Essentially, this treats each row or column as an independent data set – the results can be plotted with the other measurement location array along one axis and the power in each frequency on the other axis. This method can be useful for data sets with heterogenous dimensionality, but it is unsatisfying for data sets with homogenous dimensionality because it doesn't take advantage of the relationships between the rows and columns.

Straightforward modifications of 1-dimensional algorithms can provide the ability to identify oscillations in an arbitrary direction in a 2-dimensional data

[4]I chose a 3 km smoothing scale because that distance happened to keep the window from changing at different points on the map. Choosing another scale is possible, but it would make the window algorithm a little more complicated.

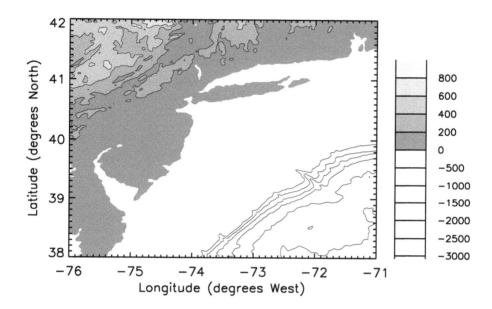

FIGURE 7.3

Contour plot of the 2-dimensional elevation data set after a 3 km radius median smoothing has been applied. The same contour-level scheme is used here
as in Figure 7.2. Below sea level, contours are drawn with solid lines every 500
m. Above sea level, filled contours are drawn every 200 m.

set. In the simplest situations, such as looking for sinusoidal oscillations in a
data set with evenly spaced elements, the frequency, power, and direction of
the waves can be measured without too much additional algorithmic complexity. To keep this discussion simple, I'm going to focus on evenly spaced
x and y measurement location arrays and data sets with no missing data.
However, the spacings of the two location arrays, given by Δx and Δy, and
their numbers of elements, given by m and n, don't have to match.

The discrete Fourier transform of an evenly spaced 2-dimensional data set
is given by:

$$
\begin{aligned}
Y_{k,l} &= \sum_{a=0}^{m-1}\sum_{b=0}^{n-1}\psi_{a,b}e^{\frac{-i2\pi ka}{m}}\,e^{\frac{-i2\pi lb}{n}} \\
&= \sum_{a=0}^{m-1}\sum_{b=0}^{n-1}\psi_{a,b}e^{\frac{-i2\pi k(x_a-x_0)}{(x_{m-1}-x_0)}}\,e^{\frac{-i2\pi l(y_b-y_0)}{(y_{n-1}-y_0)}},
\end{aligned}
\tag{7.3}
$$

where $k = 0, 1, \ldots, m - 1$ and $l = 0, 1, \ldots, n - 1$. Where I would normally use the indices i and j, I've instead used a and b to avoid confusion with the imaginary i in the exponentials. The top and the bottom forms of the equation are equivalent; which form you use depends on whether you want the measurement location arrays to be in physical or index notation. There are many similarities between the 2-dimensional form of the DFT and the 1-dimensional version in Equation 5.24 – the 2-dimensional version resembles two 1-dimensional versions, one for each of the dimensions, nested one inside the other. The two exponentials represent waves in each of the dimensions.

The relationship between the indices k and l and the frequencies of the oscillations to which they respond is the same as for the 1-dimensional case:

$$\nu_k = \begin{cases} \frac{k}{(x_{m-1} - x_0)}, & \text{for } 0 \le k \le \frac{m}{2} \\ -\frac{1}{2\Delta x} + \frac{k - m/2}{(x_{m-1} - x_0)}, & \text{for } \frac{m}{2} < k \le m - 1. \end{cases} \qquad (7.4)$$

$$\nu_l = \begin{cases} \frac{l}{(y_{n-1} - y_0)}, & \text{for } 0 \le l \le \frac{n}{2} \\ -\frac{1}{2\Delta y} + \frac{l - n/2}{(y_{n-1} - y_0)}, & \text{for } \frac{n}{2} < l \le n - 1. \end{cases} \qquad (7.5)$$

These two equations split into two cases due to the Nyquist limit, just as the 1-dimensional case did.

For each Fourier component $Y_{k,l}$, the relationship between the horizontal and vertical wave numbers k and l determines the direction of the wave associated with that component. For example, the Fourier component $(k, l) = (2,0)$ completes two oscillations over the range of the x measurement location array, but stays constant over the entire range of the y measurement location array. One way to understand the direction of the wave is to examine the wavelengths in each dimension, as depicted in Figure 7.4. Notice how the ratio of the wavelengths in the x and y directions sets the direction of the overall wave. Simple trigonometry can be used to find that the angle between the x dimension and the direction of the wave is given by $\tan^{-1}(\lambda_x/\lambda_y)$. To see how this formula is used, consider the simple case of $\lambda_x = \lambda_y$, for which the waves of the Fourier component are oriented at a 45° angle to the x-dimension.

In Figure 7.5, I've plotted the greyscale image I described in Section 7.2.2, a pattern I created by adding together two oscillations and a smaller uniformly distributed noise level. Black pixels have a value of 0 and white pixels have a value of 1, with grey pixels possessing values in between. The image is 100 pixels wide by 60 high. The reason I chose such a small image (relative to the size of most images these days) is that I wanted the 2-dimensional DFT to be easy to calculate on a personal computer. Larger images essentially require the use of the 2-dimensional FFT, which is a more complicated algorithm.

The 2-dimensional DFT of the greyscale image is shown in Figure 7.6. I've plotted the logarithm of the absolute magnitude of each complex Fourier component, normalized such that the strongest component is white and the weakest black. The axes are delineated by wavenumber in both dimensions; this means that positive frequencies are to the left (or below) the axis's midpoint,

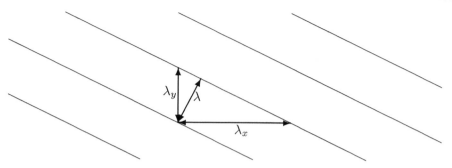

FIGURE 7.4
Illustration of the relationship between a 2-dimensional DFT component and
its 1-dimensional wavelengths. Wave crests of the component are marked with
thin diagonal lines. Wavelengths are marked with thick two-headed vectors.
The diagonal vector marked λ is the wavelength of the overall wave pattern.

and negative frequencies are to the right (or above). The peak at $(k, l) = (0, 0)$
is easy to explain – it's the DC level associated with the uniformly distributed
noise. The other two peaks, mirrored at the four combinations of positive and
negative frequencies, correspond to the two oscillations with wavenumbers
(5,2) and (2,3) from which I built this data set.

Similar to its 1-dimensional analog, the 2-dimensional inverse DFT is given
by:

$$\psi_{a,b} = \frac{1}{mn} \sum_{k=0}^{m-1} \sum_{l=0}^{n-1} Y_{k,l} e^{i2\pi ka/m} e^{i2\pi lb/n}. \tag{7.6}$$

The inverse DFT provides a means of interpolation for arbitrary $\tilde{\psi}$ at $(\tilde{x}, \tilde{y})$,
as well as a good check on your code.

The same complications to keep in mind when using 1-dimensional DFTs
can also occur in the 2-dimensional version; aliasing, ringing, and noise all
must be accounted for. As usual, don't just take the outputs of the DFT at
face value – compare the results to the original data set and make sure you
understand the origin of each peak in the spectrum.

Filtering in 2-dimensional space is similar to what was discussed in Section
5.9 for 1-dimensional data. Apply the DFT to the data, process the spectrum
with whatever filter you like, then apply the inverse DFT. Also, keep in mind
the relationship between the filters you apply in each dimension. For example,
if you apply filters of different shapes to each of the two dimensions in a
homogenous array, you had better have a very good reason.

Frequency analysis on data sets with missing elements or irregularly spacing is an advanced topic that I will not cover in this book. For further information, see Marvasti (2001) on spectral analysis of nonuniform samples for
2-dimensional signals and Bagchi and Mitra (2001)'s discussion of the Nonuni-

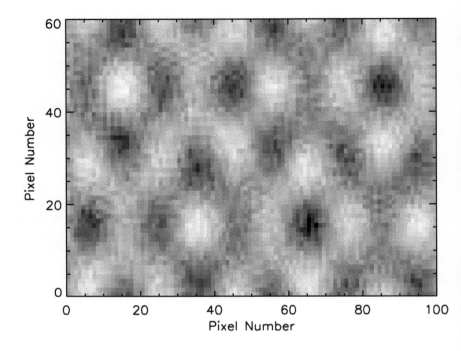

FIGURE 7.5
The greyscale image data set, created by summing two wave patterns and a
uniform noise background.

form Discrete Fourier Transform (NDFT). Since that book can be hard to find,
an alternative starting point is the brief pedagogical introduction in Hwang
et al. (2005) and the references therein.

7.4.5 Wavelet Analysis in Two Dimensions (Advanced Topic)

In Section 5.10, I discussed the application of wavelet analysis to a 1-
dimensional evenly spaced data set. The output of this technique was a heterogenous 2-dimensional data set – a structure in which the dimensions were
location and period (or frequency) and the elements were complex numbers,
which I visualized via a contour plot of the power spectrum.

Just as with 1-dimensional wavelet analysis, 2-dimensional wavelet analysis is useful for detecting localized oscillations with a variety of frequencies.
Similar to the 2-dimensional DFT, one option is to analyze each row or column
as a separate 1-dimensional data set, but this does not take full advantage of

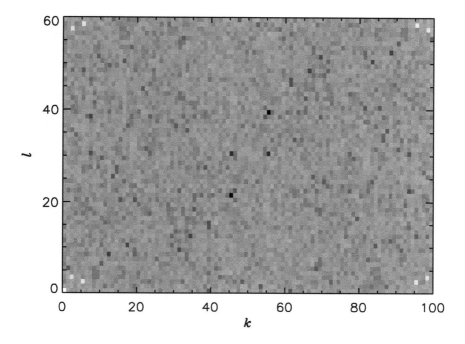

FIGURE 7.6
Plot of the logarithm of the absolute magnitude for each greyscale image DFT component. Components with strong magnitudes appear in lighter shades and those with weaker magnitudes in darker.

the relationships between a homogenous data set's elements. How does a true 2-dimensional wavelet transform work?

To apply a 2-dimensional discrete wavelet transform, it is first necessary to develop a 2-dimensional version of a mother wavelet. As with the 1-dimensional wavelet transform, there are many options for mother wavelet functions, and the function chosen should be similar in shape to the oscillations in the data. Mother wavelets in two dimensions are typically constructed by multiplying together two separate functions, one for the x dimension and a second for the y. Usually, the scale factor for the two functions is chosen to be the same – otherwise, the outputs of the discrete wavelet transform grow by an additional dimension, the computation takes longer, and the results are more difficult to interpret.

For simplicity's sake, let's imagine applying a discrete wavelet transform to a homogenous 2-dimensional data set. The response to the mother wavelet is calculated for each element in the data set (though in practice the algorithms don't operate in this fashion because it is computationally slow). The output of this process is a 3-dimensional structure – two dimensions for the position of the response and the third for the period. The elements of the output, just as in the 1-dimensional case, are complex numbers that can be converted into a real power by squaring the amplitude. This output is not easy to visualize, but it can be done with many advanced visualization software packages.

The 2-dimensional discrete wavelet transform is commonly used in image filtering, noise reduction, and compression (Kamath 2009; Press et al. 2007; Bruce and Gao 1996). It also can be applied to more traditional data analysis problems, of which Falkowski et al. (2006) is an excellent practical example.

7.5 Analogs of 1-Dimensional Techniques for Related 2-Dimensional Data Sets

Many of the techniques used to compare related 1-dimensional data sets can also be used to compare related 2-dimensional data sets. In this section, I'll discuss analogs of some of the techniques covered in Chapter 6 along with the modifications necessary to apply them to 2-dimensional data sets. In general notation, I'll examine 2-dimensional data sets $\psi_{i,j}$ and $\phi_{i,j}$, both measured at location arrays x_i and y_j.

7.5.1 Correlation Coefficients, Scatter Plots, and 2-Dimensional Data Sets

Scatter plots and correlation coefficients can be used with both homogenous and heterogenous data sets, as long as the two data sets you are comparing have been measured at some of the same locations. Since the two techniques do not depend on the actual values of the measurement location arrays, but instead only on the relationship between pairs of elements $(\psi_{i,j}, \phi_{i,j})$, there is no change in how scatter plots and correlation coefficients are constructed for 2-dimensional data sets.

If you suspect different correlations between the two data sets in different regions, it is a simple matter to construct a scatter plot or calculate a correlation coefficient for the subset of the data that requires closer investigation. Clever use of color or shading can add additional information to a scatter plot – for example, if you wanted to highlight the points in the lower left quadrant of a data set, you could shade those particular points in a different color than the rest of the data on the scatter plot. This method can be used to draw attention to a trend present in a subset of the data that doesn't appear elsewhere.

Both scatter plots and correlation coefficients can be calculated for data sets that are missing data, even if they are not missing the same data, by omitting the data pairs that are missing one or both values. However, judgment is necessary to determine whether the missing data will bias the efforts to analyze correlation. For example, if one of the data sets was measured with an instrument that broke at high temperatures, trying to find correlations between the two data sets at those temperatures would be a mistake. Take care to understand why the data is missing before accepting the outputs of the techniques.

In Figures 7.7 and 7.8 I have drawn contour plots for the climate data set's averaged temperature anomalies from the 1950s and 2000s, respectively. Both of these plots show missing data near the poles and a few other locations scattered around the globe. The two contour plots are drawn with the same contour level values. By comparing the two plots, it is easy to see that the 2000s plot has larger temperature anomalies than the 1950s plot.

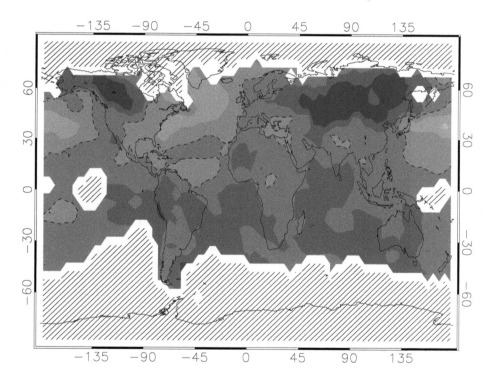

FIGURE 7.7

Contour plot of the averaged temperature anomalies for the 1950s. Degrees of latitude and longitude are marked on the axes, and a map of the global landmasses is drawn in the background with a solid line. A dashed contour appears at the 0°C temperature anomaly level. Contour levels are separated by 0.4°C, with lighter contours denoting positive anomalies and darker contours negative ones. Portions of the map marked with diagonal hashmarks are missing data.

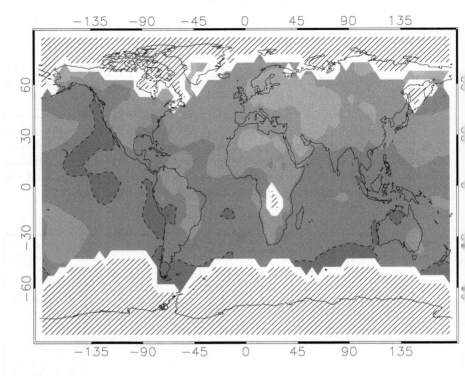

FIGURE 7.8
Contour plot of the averaged temperature anomalies for the 2000s. Degrees
of latitude and longitude are marked on the axes, and a map of the global
landmasses is drawn in the background with a solid line. A dashed contour
appears at the 0°C temperature anomaly level. Contour levels are separated by
0.4°C, with lighter contours denoting positive anomalies and darker contours
negative ones. Contours are shaded identically to those in Figure 7.7. Portions
of the map marked with diagonal hashmarks are missing data.

Figure 7.9 depicts the scatter plot for the relationship between the 1950s
temperature anomalies and the 2000s temperature anomalies for locations
where neither decade is missing data. The points appear clustered in a flat
shape displaced about 0.5°C above zero on the vertical axis and slightly left of
zero on the horizontal axis, consistent with a global increase in the temperature anomaly over that ≈ 50 year time period. Similarly, the median temperature anomaly increased from −0.27°C to 0.36°C over that time period (again
measured at locations where neither decade is missing data). There does not
appear to be a simple correlation between grid points that were hotter than
average in the 1950s data and those that were hotter or cooler than average
in the 2000s data. The Pearson correlation coefficient, calculated to be 0.08,
supports the lack of this relationship.

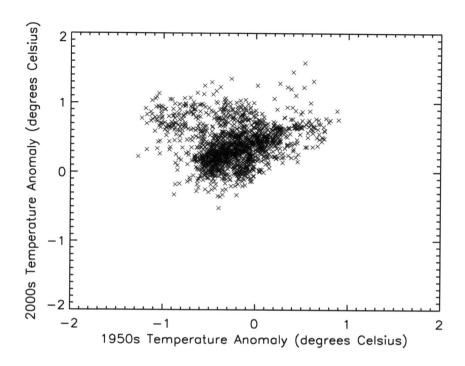

FIGURE 7.9

Scatter plot comparing the climate data set's temperature anomalies from the 1950s and 2000s. Each 'x' represents the pair $(\psi_{i,j}, \phi_{i,j})$ associated with a location (x_i, y_j).

7.5.2 Ratios of 2-Dimensional Data Sets

Ratios are a helpful technique to use in combination with scatter plots and correlation coefficients for data sets measured with ratio scales on the same measurement location arrays. Ratios can be used to identify regions of the data wherein the relationship between the two multidimensional arrays is different than for the rest of the data. There are no major differences between how they are employed with 2-dimensional data compared to 1-dimensional data. Of course, the ratio of two 2-dimensional data sets is a 2-dimensional object in itself, requiring mapping and visualization techniques to properly display and analyze.

Since temperature anomalies from the climate data set are measured with an interval rather than a ratio scale, it is not appropriate to take the ratio of the two decadal average structures. Instead, an analyst could construct a quantity similar to a ratio by subtracting one of the structures from the other.

7.5.3   2-Dimensional Cross- and Autocorrelation Functions

Calculation of the 2-dimensional cross-correlation function follows straight-forwardly from the 1-dimensional discussion (Section 6.7). Adding an extra dimension of data simply requires the displacement of the data to be 2-dimensional. This means that the outputs of a 2-dimensional cross-correlation function are 2-dimensional structures themselves, with the measurement location arrays marking the displacement in the two directions, and the elements containing the response function.

The 2-dimensional version of the normalized cross-correlation function is given by:

$$\hat{X}_{a,b} = \frac{\sum_{i=0}^{m-1}\sum_{j=0}^{n-1}\left(\phi_{i+a,j+b}-\mu_\phi\right)\left(\psi_{i,j}-\mu_\psi\right)}{\sqrt{\sum_{i=0}^{m-1}\sum_{j=0}^{n-1}\left(\phi_{i,j}-\mu_\phi\right)^2\sum_{i=0}^{m-1}\sum_{j=0}^{n-1}\left(\psi_{i,j}-\mu_\psi\right)^2}} \tag{7.7}$$

where the means of the two data sets are given by μ_ψ and μ_ϕ and the combination (a,b) describes the offset in the (x,y) directions in index form.

This version of the cross-correlation function technique is intended to be used when there is no missing data. Extra care is required when data is missing to ensure that the cross-correlation function is not biased when the missing data points overlap, which reduces the total number of pairs with missing data.

The autocorrelation function is useful in finding repetitions within a single data set. The same caveats that apply to the 1-dimensional version apply to the 2-dimensional one.

7.6   Higher-Dimensional Data Sets

I won't discuss them in depth in this book, but higher dimensional structures are sensible choices for some data sets. Here are a few examples just to get you thinking about how they could be structured:

- The 3-dimensional mass density of the Galaxy;

- The maximum daily temperature at a grid of locations around the Earth, with the first location array corresponding to latitude, the second to longitude, and the third to the day of the year.

- A data structure in which the first three dimensions spatially map a room, the fourth is time, and the elements themselves are air densities measured after a space heater has been turned on.

Additionally, the climate data set described in Section 7.2.3 is actually a 3-dimensional heterogenous data set, with two spatial dimensions (longitude and

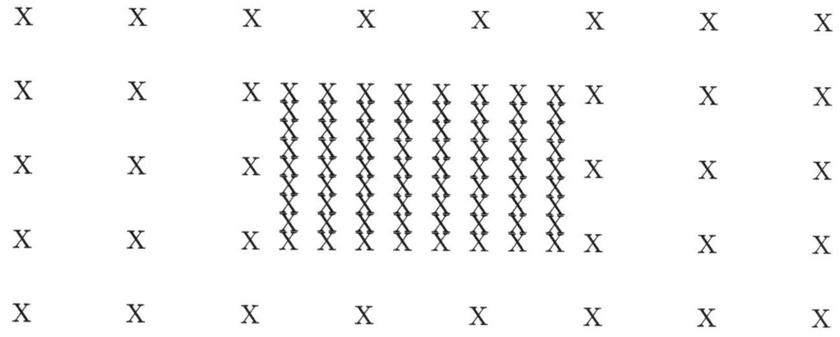

X   X   X   X   X   X   X   X

FIGURE 7.10
Example illustration of element locations in a 2-dimensional mesh.

latitude) and one temporal dimension. If you'd like to practice some techniques on a higher dimensional data set, the climate data set is a good place to start.

Techniques that rely on the homogeneity of a data set's dimensions can sometimes be applied to subsets of a heterogenous data set. For example, consider the grid of temperature measurements measured with one temporal and two spatial dimensions. By examining a single moment in time, a 2-dimensional homogenous data set can be extracted from a 3-dimensional heterogenous one.

Adding more dimensions isn't the only way to increase the complexity of data structures. For example, some advanced analytic applications use a specialized type of structure called a <u>mesh</u>. A mesh is a data structure in which the spacing of the elements in one or more of the location array varies depending on the other location arrays. In other words, some region of the structure has a different, usually smaller, spacing between the elements (Figure 7.10). The location arrays in a mesh cannot be written simply as (x_i, y_j) because the number of elements varies by row. Meshes are used in applications where higher resolution is necessary for some portion of the structure, but using that high resolution everywhere in the array is prohibitive for some reason (computational time, storage space, etc.). Meshes can be built with multiple levels of finer granularity, and can even be adjusted in the middle of a simulation to model with greater resolution areas of the data where something interesting is happening (Berger and Colella 1989).[5]

---

[5]Note that in 1-dimensional data sets, meshes can be handled by an unevenly spaced location array – it's only when the number of dimensions grows that the mesh formalism becomes convenient.

7.7   Exercises

1. Practice restructuring by taking the elevation map and transforming it from the original three grouped lists to a 2-dimensional data set. Then transform it back again.

2. Construct a contour map for a region of the ETOPO1 data set near your hometown. Apply some simple analytic techniques to this data set, such as mean, median, and standard deviation. Construct a histogram of the elements.

3. Apply a running mean filter to the elevation map. Compare its effects to the median filter, first by eye, and second by taking the ratio of the two smoothed maps. Why are or why aren't they different?

4. Calculate the Fourier power spectrum for the greyscale image. Confirm that Parseval's Theorem holds for this data set, i.e., that the total power is identical in the original image and in the Fourier transformed data.

5. Apply a bandpass filter to the greyscale image that passes a range of frequencies near the two wavelike functions. Plot the image that results after the filter has been applied.

6. Calculate the Kendall τ for the 1950s and 2000s temperature anomalies from the climate data set.

7. Construct a scatter plot for 1940s and 1990s decadal averages from the climate data set.

8

Unstructured Data Sets

8.1 Introduction

I'm sure there are many readers that, to this point, have found this book frustrating because their data lack the structure of the data sets I've described, making the techniques I have discussed difficult or impossible to apply. Fortunately, like structured data, unstructured data is also amenable to analysis. In fact, the analysis of unstructured data is a subject all analysts should be at least somewhat familiar with because it is applicable in many surprising situations. However, in an unstructured data set, the elements' lack of organization can make applying even the simplest of techniques into a real struggle that requires some creative thinking. For these reasons, the discussion in this chapter is an introduction to analyzing unstructured data.

I won't say unstructured data is the most difficult type of data to analyze, but it may very well be the strangest. An important idea to remember is, as I discussed in Section 1.3.1, the data are not trying to tell you anything. This is particularly relevant to unstructured data sets, because it is difficult for analysts trained on structured data to develop an intuition for what questions can or cannot be answered with unstructured data. Not all the analytic questions you are interested in are going to be answered by a single data set.

Unstructured data comes in a variety of forms, so some examples will help clarify the term's meaning. Examples of unstructured data include:

- The collected plays of William Shakespeare,

- A photograph of a flock of birds flying against a blue sky, and

- All of the articles published on a news website on a particular day.

These are just a few examples, but what constitutes unstructured data is very broad and, as usual, is subject to disagreement amongst different analysts. In my view, this disagreement results from the difference in perspectives with which analysts approach a data set – let me explain.

What characteristics make a particular data set unstructured? After all, if you look at the examples above from a particular point of view, they have plenty of structure. For example, a digital version of the photograph could be structured as a 2-dimensional evenly spaced data set with each element encoded with a color, like the greyscale image I introduced in Chapter 7. Similarly, Shakespeare's plays could be encoded into a 1-dimensional data set where each letter, number, and punctuation mark is individually represented as an element. For both of these examples, applying certain analytical techniques to the data sets is not difficult at all. So why do most analysts think of them as unstructured data sets?

The crux of the issue is that the structure that does exist is not useful for answering the questions that are most interesting to analysts. For the digital photograph, the existing structure doesn't provide an easy way to count how many birds are in the picture. For Shakespeare's plays, the structure makes it difficult to determine the meaning of each sentence. In other words, whether a data set is unstructured is often in the eye of the beholder – <u>unstructured data</u> is data whose existing structure doesn't provide a straightforward path to answering the analytic questions of interest. Two analysts using the same data set to answer different questions of interest could disagree on whether or not the data set was unstructured, and they would both be correct.

There are essentially two methods for analyzing unstructured data sets: extracting a structured data set from the unstructured one and finding a creative workaround that answers the question of interest. The next section discusses the first of these methods, followed by a deeper look at the analysis of unstructured text, and the final section of the chapter covers Bayesian techniques, which are useful for creatively analyzing even the most discouraging unstructured data set.

8.2 Extracting Structure from Unstructured Data

Much of the effort that goes into analyzing unstructured data is expended to transform the data into a form where analytic techniques designed for structured data can be applied.[1] For example, imagine a piece of software that extracts the position of each bird in the photograph of the flock, perhaps also returning the bird's size and color. The output of this software would be a group of related lists – a separate list for the birds' horizontal positions, vertical positions, sizes, and colors. These outputs, much like the measurements from a scientific instrument, would have associated error and uncertainty (as discussed in Chapter 4). The software that performed this extraction would need to be sophisticated to distinguish birds from clouds, airplanes, and other objects. It's clearly not a trivial program to build.

In this example, the software is used on unstructured data to extract a

---

[1] In Sections 8.2 and 8.3, I will deviate from my usual strategy of presenting solved examples because the computational algorithms involved in extracting structure from unstructured data are so complicated. Instead, I will set up simple examples to illustrate the complexity of the subject matter and to make you aware of the relevant subtleties.

structured data set. In general, this process is known as preprocessing – it's the work that is done to the original "raw" data to prepare it for analysis. For unstructured data, preprocessing restructures the data into a form in which it can be interrogated to answer the analytic questions of interest. However, preprocessing isn't limited to unstructured data, because many analysts also use the term to describe any preparatory work to ready data for an analytic technique to be applied (as I did for the climate data set in Section 7.2.3). For example, preprocessing can include cleaning data, reducing the data's size, combining data from different instruments, applying normalization factors to the data, and many other actions.

Metadata is a key concept in extracting structure from unstructured data, but analysts don't use the term consistently. The most common definition of metadata is "data about data," but while that phrase is a good summary of the concept, in my view it is too vague to be useful. Instead, I'll use a modified version of one of the definitions from the literature to describe two simple categories of metadata that are intended to be mutually exclusive and collectively exhaustive: control and guide (Bretherton and Singley 1994). Control metadata refers to information about the structure of the data set, whereas guide metadata pertains to the content of the data set.[2]

Control metadata accompanies many data sets, either embedded in the digital files or in the accompanying written documentation. For example, many digital cameras will attach date, time, size, and location information to the image file. This information is later referenced by software that catalogs and displays the images and is used to automatically build albums of related pictures. The column headings in a spreadsheet can also be considered control metadata (Inmon and Nesavich 2008), as can a website describing the structure of a data set (as the ETOPO1 website does for the elevation data set).

Guide metadata, on the other hand, depends on the content of the unstructured data set, and therefore must be determined after the data has been gathered. The kind of guide metadata you build also depends on the question in which you are interested. Examples of guide metadata include the times of the goals accompanying a video of a soccer match, the page and line numbers of all the section headings in a book, and the recipients' addresses in a database of emails. Of the two types of metadata, guide metadata is more relevant to analyzing unstructured data sets; in fact, the analysis of an unstructured data set is often accomplished via the application of structured techniques to the guide metadata rather than directly to the unstructured data.

Part of the difficulty in analyzing an unstructured data set lies in actually determining the guide metadata. Methods for determining metadata depend on both the data set's type and the questions the analyst would to like answer with the metadata. For example, I might construct different guide metadata

---

[2]Guide metadata was originally defined to be written in natural language for humans and not intended for computer interpretation, but I'm relaxing that restriction.

for an application that searched documents than for one that classified documents. Automated applications for determining guide metadata will be very different for text, video, audio, and other types of unstructured data.

There isn't necessarily a relationship between the size of the unstructured data set and the guide metadata that is drawn from it. For example, consider the digital photograph of the flock of birds. If the picture is high resolution, the image file could be many megabytes in size, containing millions or more elements. If the metadata extracted from the image is simply the horizontal and vertical positions of each bird, there are only two elements per bird. If there are only ten birds in the image, that's twenty elements, a structured data set many orders of magnitude smaller than the unstructured image data.

For unstructured data sets for which the information to be extracted is small, it is possible to generate the guide metadata by hand. Consider the time it would take to complete the extraction by hand and the gain or loss in time and precision that would result from automating the procedure. Typically analysts prefer the guide metadata to be constructed in an automated fashion because creating it manually is tedious, but sometimes building a software solution is so difficult that extracting guide metadata by hand is the only reasonable path.

8.3   Text Mining

To illustrate the analysis of unstructured data in more depth, let's discuss how to analyze text data. Modern society produces a tremendous volume of text as part of everyday activities, such as emails, web pages, blog posts, social media status updates, and text messages. The information contained in this data can be used for many purposes, such as news and marketing, but it isn't obvious how that information can be extracted. The process of extracting information from unstructured text is known as text mining. The discussion in this section is drawn primarily from the excellent introduction by Weiss et al. (2010).

An unstructured text data set is a collection of documents. Depending on the question you are trying to answer and the data at hand, these collections can be extremely broad, such as all the websites on the Internet, or more narrow, such as a particular company's emails from the last year. The documents themselves can take many forms, such as emails, reports, websites, and contracts, or some mixture of types. Documents can have a variety of components, including (but not limited to) words, numbers, and punctuation. They may also have more specialized components, such as email addresses, phone numbers, and web addresses, depending on the type of document. They may follow a particular format, such as the to, from, and subject headers on top of an email, or they may be unformatted. Text mining is extremely flexible and can accommodate all these types of input.

Useful guide metadata can be created for a collection of documents based solely on the frequency and position of words, without having to understand the meaning of a document or its constituent sentences. This is convenient, as it's much easier to design software to extract the frequencies and positions of words than it is to write an algorithm to understand the vagaries of language.[3] In fact, many problems can be solved by analyzing this type of guide metadata, such as:

- Classification – Is this email spam or legitimate?

- Clustering – Take this collection of reports and sort it into piles of similar documents.

- Search – Find all the contracts our company has signed involving a certain piece of hardware.

- Information extraction – Populate this spreadsheet with the values of all the contracts the company has signed.

By looking more closely at the search problem, I will illustrate some of the complexities involved in analyzing unstructured text data. Let's imagine I've used a piece of software to build guide metadata for a collection of documents. The guide metadata contains counts of the number of times each word appears in each document. Because exact match searches are very sensitive to the choice of word, if I were to search for one particular word, I might not get the results I expect. For example, if I'm doing research on how the U.S. President is covered by a particular news organization and I extract all the occurrences of the word "President" in an unstructured data set of news articles, I have to worry about my sample being contaminated by presidents of corporations, universities, and other countries. Furthermore, my simple search won't find articles in which the President is referred to by his alternate title, the Commander-in-Chief, or by his last name. Similar issues would arise if I were to search for the word "cat" – shouldn't "cats" also be returned? Shouldn't "feline"?

Text mining techniques have been developed to handle all of these different cases. One technique, known as stemming, converts words to a standard form when the guide metadata is built. For example, inflectional stemming can be used to remove the pluralization from words and the tense from conjugated verbs. Synonym dictionaries can be used to connect related words, such as "cat" and "feline." More complex techniques, such as those that depend on a word's part of speech or its presence in a multiword phrase have also been developed.

There are two types of errors than can occur in a search intended to identify which documents a word appears in: a document that should have been

---

[3] Using a computer to determine the meaning of an inputted text document is the purview of the field of natural language processing.

a match can be missed or a document that should not have been a match can be mistakenly included. These are known, respectively, as errors of <u>omission</u> and <u>commission</u>. These types of errors are not limited to text mining; analogous behavior in the image of the flock of birds is easy to imagine. You may miss a bird because your view is partially blocked by one in front of it (omission), or your software may mistake a jet plane in the background as a bird (commission).

Clearly, building a piece of software to extract metadata from unstructured data is fraught with subtlety and complexity. It also requires extensive effort, attention to detail, and computational firepower. When these resources aren't available, an alternative technique for analyzing unstructured data would be useful.

8.4   Bayesian Techniques

From the previous sections in this chapter, you've gotten a sense for how to analyze unstructured data sets by extracting guide metadata. This section demonstrates an alternate technique for analyzing unstructured data: Bayesian reasoning. Bayesian techniques are versatile and can also be used on structured data.

Before applying these techniques to unstructured data, I will first explain some background on probability. The connection to analyzing unstructured data will soon become apparent.

8.4.1   Bayes' Rule

In science and mathematics, there are occasionally equations that are simple in form, but convey powerful and sophisticated ideas. <u>Bayes' rule</u> is one of these equations. Essentially, it is a way of relating the probabilities of occurrence of two events, call them A and B. Think of A and B as any events you can imagine: a coin flip turning up heads, a six-sided die turning up three, the Yankees winning the World Series this year, or my car breaking down on its next trip.

I will write the probabilities of A and B occurring as $p(A)$ and $p(B)$. In addition to these simple probabilities, there also exist <u>conditional probabilities</u> of the events: $p(A|B)$, which is the probability of A occurring given B has already occurred, and $p(B|A)$, which is the probability of B occurring given A has occurred. These probabilities are clearly more interesting when A and B are not independent random events; in other words, when $p(A|B) \neq p(A)$. For example, imagine you were trying to estimate the probability of my car breaking down on its next trip to the grocery store (call this event A). You'd likely assign that event a small probability, because most cars are pretty reliable. Now, let's say you observe event B: the check engine dashboard light is on. This certainly influences your estimate of event A happening; in other words, my car breaking down and the check engine light being illuminated are not independent random events.

Bayes' rule mathematically relates these quantities:

$$\frac{p(A|B)}{p(A)} = \frac{p(B|A)}{p(B)}. \tag{8.1}$$

If you think about A and B as abstract events, there isn't anything especially profound about Bayes' rule. Let's examine this equation in a limiting case to demonstrate. Assume that if A occurs, then B is certain to occur, and vice versa; $p(B|A) = p(A|B) = 1$. Then, Equation 8.1 can be rearranged to read $p(A) = p(B)$; this clearly must be true if A and B are inextricably linked by their conditional probabilities.

To be especially pedantic, I could have written all of the probabilities in Equation 8.1 as conditional on the information "I" available at the time. In other words, by writing $p(A|I)$ I'm specifying my beliefs regarding the probability of event A occurring given the information currently available. For example, if I were estimating the conditional probability $p(\text{car breakdown}|\text{check engine light on}, I)$, the probability incorporates, via "I," the sum of my experience with cars and indicator lights. Writing out "I" explicitly might seem a little strange; after all, am I not always assuming the information I already have? Indeed, but making this assumption explicit is important because it makes it clear that I will need to reevaluate the probabilities whenever new background information comes to light.

The discussion up to this point has been quite abstract, so let's clarify with an example. Let's say I am trying to determine whether a given manufactured part is faulty. There is a test to check each part, but that test sometimes indicates that a correctly made part is faulty; this situation is called a false positive. So, if a part fails the test, what is the probability that it is actually faulty (in other words, a true positive)?

Bayes' rule is a great way to approach this problem. Let's make the assumption that I understand the test pretty well because I've run a large number of parts through it in the past. Of those previous parts, 99% of the correctly made parts passed the test with the remainder failing; this implies $p(\text{failed test}|\text{correct part}, I) = 0.01$. Also, 97% of the faulty parts failed the test with the remainder passing; this means $p(\text{failed test}|\text{faulty part}, I) = 0.97$. It seems like the test is pretty accurate with those probabilities, but let's see how the numbers play out.

First, I need some more information about the set of parts as a whole. Again, I'll use background knowledge to say that only about 1% of constructed parts are actually faulty; this quantity is called the prior probability and is written $p(\text{faulty part}|I) = 0.01$. It's called a prior probability because it's my estimate of whether the part is faulty before I have any specific information about that particular part. I'd like to calculate the probability that a part which fails the test actually is faulty, written mathematically as p(faulty part|failed test, I). In Bayesian analysis, this quantity is called the posterior probability because it incorporates the new information from the test and represents the analyst's beliefs after that information has been assimilated.

Let's plug some of these numbers into Bayes' rule. First, I can calculate the overall chance of the test returning a faulty result from my knowledge of the test's accuracy and of the production rate of faulty parts (these are quantities I've already described); this yields p(failed test|I) ~ 0.02. Half of that rate is from false positives, and the other half is from true positives (confirm this yourself). So, in this simple example:

$$p(\text{faulty part|failed test, I}) = \frac{p(\text{failed test|faulty part, I})}{p(\text{failed test|I})}p(\text{faulty part|I})$$

$$= \frac{0.97}{0.02}(0.01) \sim \frac{1}{2} \qquad (8.2)$$

So, even when the test returns faulty, the part is only actually faulty half the time! Such is the influence of the false positives.

What are the chances that a part that passes the test is actually faulty? Bayes' rule is perfect for this question too; I just need to find p(faulty part|passed test, I). I already know p(faulty part|I), and p(passed test|I) ~ 0.98 from my earlier calculation of p(failed test|I). Using knowledge of the conditional probabilities, I find a fairly small chance of a faulty part given a passed test, p(faulty part|passed test, I) ~ 0.0003. These are called false negatives.

How much did my choice of prior probabilities influence the result? Solving Bayes' rule for a "worst-case scenario" of p(faulty part|I) $= 0.10$, which increases p(failed test|I) to 0.106, yields p(faulty part|failed test, I) $= 0.92$. In other words, when there are more faulty parts made, the chance that any given faulty test result actually corresponds to a faulty part increases. This is a fairly big change in the true positive rate, but I did increase the fault rate by an order of magnitude.

How does Bayes' rule relate to the analysis of unstructured data? In fact, many problems reliant on unstructured data can be solved with Bayesian logic. Consider the classification of an email as spam or legitimate – this is a categorical measurement. One way to judge which category an email falls into is by connecting it using Bayes' rule to the probability that it comes from a known sender. Then, it may be possible to develop a spam filter without searching through the text of previous emails.

For the discussion that follows in the next section, it is important to note that the mathematical relationships described here are bidirectional. For example, if you knew how a given part tested, you could calculate the probability of it being faulty, and if you knew whether a given part was faulty, you could calculate the probabilities of how it would test. However, unlike the mathematical relationships, the causal relationship between the variables only goes one direction: the state of the part influences the outcome of the test. In graphical form, I can represent this as a simple <u>causal network</u>:

Each box in a causal network is called a <u>node</u>. Each node can take one or more states, often described by categorical variables, such as a part being correctly made or faulty or a test result reporting pass or fail. There isn't much meat to this particular diagram, but causal networks become increasingly useful the more complicated a problem gets.

8.4.2   Bayesian Belief Networks

Often, when trying to make a measurement on a categorical scale, there are multiple pieces of evidence that should influence that classification. In the previous section, I probabilistically classified a part as correctly made or faulty based on the results of a single test, but what if there were more pieces of evidence or information that would help me make that classification? A <u>Bayesian belief network</u>, also known as a Bayesian decision network or a relevance diagram, is a technique used to connect multiple pieces of information and calculate the influence of known information on the other nodes in a causal network. In this section, I will demonstrate how additional information can be incorporated by a Bayesian belief network using two simple examples.

For the first example, let's add another node to the simple causal network from the previous section. Suppose the manufacturing machine is more likely to correctly shape the part when the machine's temperature is kept lower than a certain operating threshold. When the temperature is lower than this threshold, about 1% of the parts that are manufactured will be faulty, as I've described above. However, when the machine's temperature exceeds the threshold, the rate of faulty parts increases to 10%. Graphically, I can represent this situation as a new node on the causal network:

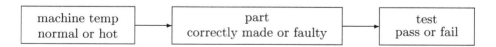

In plain language, this diagram means that the machine temperature influences whether the part is manufactured correctly, which influences the outcome of the test. How can I represent these relations mathematically?

If I knew the machine temperature at the time when any given part was made, I could just use the prior probability that applies to that situation for p(faulty part|I) in Equation 8.2 to calculate the probability of each test outcome. The more complicated (but still solvable) problem occurs when all the manufactured parts are mixed together and I don't know the machine temperature when any single part was made, just an estimate of how often the machine temperature is too high. Let's investigate that case.

Helpfully, the equations that describe the relationship between the state of the part and the outcome of the test are unchanged from the ones I described in the previous section (though the values of some of the variables have changed). I only need to write equations describing the new relationship I have introduced. Applying Bayes' rule a second time yields:

$$\frac{p(\text{machine hot}|\text{faulty part}, \text{I})}{p(\text{machine hot}|\text{I})} = \frac{p(\text{faulty part}|\text{machine hot}, \text{I})}{p(\text{faulty part}|\text{I})}. \tag{8.3}$$

The piece of information missing here is a rough estimate (a prior probability) of how often the machine is hot, which can be used with the conditional probabilities of faulty part manufacturing given the state of the machine to estimate the new p(faulty part|I). I can then use that probability in Equation 8.2 along with an adjusted p(failed test|I) to calculate the new p(faulty part|failed test,I).

This first example is a simple illustration of how a Bayesian belief network works. If the temperature of the machine is known, the other nodes follow the equations pertaining to that state of the machine. If the temperature of the machine is unknown, Bayes' rule can still be used to calculate the probabilities of the other nodes. Similarly, if the test result for a particular part was known to be "failed," I could solve the equations to yield an updated probability for the temperature of the machine when that part was manufactured. Information flows both with and against the direction of causality.

For the second example, let's make the problem even more complicated by moving away from a linear causal network. Another way I can add information to the original causal network is to introduce a separate and independent test to determine if the part is correctly made. Call this the expensive test; if the first cheaper test is a relatively low precision measurement of the color of the part, this second more expensive test could measure the part's volume or some other quantity to a high precision. This configuration implies the following causal network:

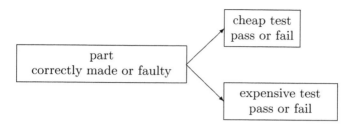

It is easy to imagine an analyst needing to answer a variety of questions for the testing scenario I've described; for example:

- If a part fails the cheap test but passes the expensive test, what is the probability that it is correctly made?

- If a part fails the cheap test, what is the probability that it will pass the expensive test?

- If a part is faulty, what is the probability that it would pass both tests?

Bayesian belief networks use the algebraic connections between the nodes given by Bayes' rule to answer questions such as those I've listed. These questions (and their related variants) stipulate the value of some of the nodes in the network, then ask about the repercussions for the other nodes. The questions often don't proceed in the direction of causality, such as a question that begins by stipulating a test result. Clearly, that information influences the analyst's belief regarding whether or not the part is faulty – Bayesian belief networks connect the acquired test evidence to the analyst's beliefs regarding whether the part is faulty. Again, unlike in causal networks, information (and algebra) flows in all directions in a Bayesian belief network.

Let's quickly explore how to answer the first example question using a Bayesian belief network. I need to solve for the probability that the part is correctly made given that it failed the cheap test but passed the expensive test: $p(\text{correct part}|\text{fail cheap, pass expensive, I})$. Let's call this p_1 for short. By using Bayes' rule, I can state:

$$
\begin{aligned}
p_1 &= \frac{p(\text{fail cheap, pass expensive}|\text{correct part, I})p(\text{correct part}|\text{I})}{p(\text{fail cheap, pass expensive}|\text{I})} \\
&= p(\text{fail cheap}|\text{correct part, I})p(\text{pass expensive}|\text{correct part, I}) \\
&\quad \times \frac{p(\text{correct part}|\text{I})}{p(\text{fail cheap, pass expensive}|\text{I})},
\end{aligned}
\tag{8.4}
$$

where the second line of the equation takes advantage of the conditional independence of the two tests. Let's call the denominator of the fraction p_2 for short; it can be reduced to known quantities using the law of total probability:

$$p_2 = \left[p(\text{fail cheap, pass expensive}|\text{correct part}, I)p(\text{correct part}|I) \right]$$
$$+ \left[p(\text{fail cheap, pass expensive}|\text{faulty part}, I) \right.$$
$$\left. \times p(\text{faulty part}|I) \right], \tag{8.5}$$

which can be further simplified using conditional independence. This reduces p_1 to a combination of known quantities and probabilities that can be straightforwardly calculated.

Bayesian belief networks can be much more complex than the three node varieties I've described here in these two examples. Though there are some subtleties to solving the networks based on which nodes are causally connected, the principles of the solution follow simply from Bayes' rule.

Bayesian belief networks can also be used to determine the value of additional information, which is a difficult problem to solve using other techniques. Consider a part that fails the cheaper test; this result could be a false positive. Is it worth it to perform the more expensive test on the part to see if the result is confirmed? A Bayesian decision network can tell you how much more certain you would be that a part was faulty if it failed both tests; this is the value of the new piece of information that results from the expensive test. Whether it is worth it to run the extra test depends on the cost of the expensive test and the relative accuracy of the two tests.

There are some general rules that govern constructing these networks. Most importantly, causal networks can't contain any circular loops, because a loop would contradict the rules of causality. Additionally, as the causal network for a problem gets more and more complex (and especially as it breaks further and further from linearity), more and more probabilities are necessary to populate the Bayesian belief network.

It's easy to see how to apply Bayesian belief networks to problems with unstructured data. Imagine again trying to build an email spam filter – several pieces of information could influence the decision of whether a given email is spam or legitimate. In addition to whether the sender is known, the time the email was sent and the location the IP address corresponds to could influence the decision. This information influences the chance of the email being spam, and could be built into a causal network resembling the one with the cheap and expensive tests.

The techniques described in this section can be straightforwardly generalized to categorical scales that have more than two states and to networks of arbitrary size. The algebra involved gets tedious fairly quickly, but there are computer programs like Netica available to organize and solve complex networks. Software written specifically to solve these networks often will have an easy-to-use interface allowing you to select the state of the evidence for any particular iteration of your problem, which can be helpful if you have less technical people trying to use the network. Or, if you're ambitious, you can code up a solution yourself. More information about Bayesian belief networks can be found in Schachter (2007) and the freely downloadable Krieg (2001), and there is a collection of examples in Pourret et al. (2008).

8.4.3 Bayesian Statistics

Let's step away from the analysis of unstructured data and look at a special case of Bayes' rule where A and B are defined more precisely (Sivia and Skilling 2006). Determining the truth of a hypothesis is a common task in analytics for all types of data sets. Let's call a hypothesis I am investigating event A: if the hypothesis is true then A has occurred, and if it is false then A has not occurred. These are the only two possible outcomes, in other words $p(\text{true hypothesis}|\text{I}) + p(\text{false hypothesis}|\text{I}) = 1$. I've essentially assigned a categorical scale measurement to the truth of the hypothesis, with the two categories mutually exclusive and collectively exhaustive.

Next, let's say event B is the data I have observed in the course of an experiment. With these substitutions, I can rewrite Bayes' rule (Equation 8.1) as:

$$\frac{p(\text{true hypothesis}|\text{data},\text{I})}{p(\text{true hypothesis}|\text{I})} = \frac{p(\text{data}|\text{true hypothesis},\text{I})}{p(\text{data}|\text{I})}. \tag{8.6}$$

Looked at through this lens, analytics is an attempt to determine $p(\text{true hypothesis}|\text{data},\text{I})$, that is, the posterior probability that a hypothesis is true given the observed data and the background information. When Bayes' rule is used to test the truth of a hypothesis in this way, the process is known as <u>Bayesian statistics</u>.

Note how different Bayesian statistics are from the other techniques I have presented in this book. Through the prior probability $p(\text{true hypothesis}|\text{I})$, Bayesian statistics incorporate all of the background information at the analyst's disposal. Contrast this with the significance levels I calculated in previous chapters (such as Section 3.7); these techniques fall under <u>frequentist statistics</u>. In frequentist statistics, each data set stands or falls on its own and prior knowledge is deliberately excluded.

For these reasons, Bayesian statistics are often criticized for being more "subjective" than frequentist statistics. This criticism stems from the influence of the prior probability $p(\text{true hypothesis}|\text{I})$ in Equation 8.6. Essentially, the analyst's beliefs regarding the truth of the hypothesis before the data are collected influence his or her beliefs of the hypothesis's truth after everything is said and done. Frequentists view this as bias entering the analysis, but Bayesians look at it as zeroing in on the right answer. After all, it's not like an analyst discovers the world anew each day.

Modern statistical analysis is shifting from the frequentist viewpoint to the Bayesian perspective. However, in practice, analysts must know how to use both types of techniques as both are commonly used. If you'd like to learn more about how both types of statistics reflect events in the real world, see the excellent (and nontechnical) discussion in Silver (2012).

8.5 Exercises

1. Imagine you have an unstructured data set that consists of family snapshots. Name three examples of guide metadata you could extract from this data.

2. Do the same for all of the papers you wrote in high school.

3. Construct a Bayesian decision network that categorizes the text documents on your computer into a set of mutually exclusive categories (financial, entertainment, work, etc.). Incorporate into the network a few simple tests that you could perform on the files. Estimate some of the conditional and prior probabilities based on your knowledge of the files, and use that information to calculate the probabilities you couldn't estimate. Think of three questions you could use this network to answer, similar to the questions in Section 8.4.2.

9

Prescriptive Decision Analysis

9.1 Introduction

In this chapter, I will describe a few techniques from prescriptive decision analysis,[1] a subject that is part of prescriptive analytics. As I discussed in Chapter 1, the larger subject of prescriptive analytics pertains to the use of data to inform action. Within that space, prescriptive decision analysis provides models to connect the world of decision making with the realm of data. It's an essential topic for analysts who want their work to influence decisions made by real people.

Though it had precursors in earlier work by Bernoulli, Laplace, and others, decision analysis really got started with the application of operations research techniques to military problems in the 1940s. The first practitioners were also influenced by early game theory, managerial economics, and probability theory (Raiffa 2007). Decision analysis's current academic home is typically in business schools and operations research departments, with some extensions into economics and psychology programs. Despite its academic, military, and governmental origins, prescriptive decision analysis is useful in a wide variety of analytic contexts, especially the business world and personal decision making (see Hammond et al. (1999) for a layman's discussion).

For readers that come to this book from a scientific or academic background, it will be easier to see the applicability of prescriptive decision analysis to your work if you keep some relevant examples in mind. Think about applying these techniques to your decisions of what research question to investigate next (see Section 1.3.1) or which piece of lab equipment to procure with a limited budget.

My intent in this chapter is to give you a flavor for some of the various methodologies that are part of prescriptive decision analysis. This chapter is not meant to be comprehensive, but it will give you some intuition for how to connect your analysis to decisions as well as suggestions for where to turn for more detail. A more technical introduction to the subject can be found in *Making Hard Decisions* (Clemen and Reilly 2001).

---

[1] Other types of decision analysis include normative decision analysis, which characterizes rational and logical decisions, and descriptive (or behavioral) decision analysis, which studies how real people behave.

Let's begin by describing the basic structure of a decision utilizing a few common terms of art. First, the term <u>decision</u> is defined broadly, referring to a choice, with some impact, that one or more people make. The <u>decision maker</u> is the entity with the authority to make the choice and commit to the decision. <u>Alternatives</u> are the options that the decision maker chooses between in the course of making the decision. <u>Outcomes</u>, also known as payoffs or consequences, are the effects that result from the alternatives under consideration. So, in very broad strokes, my basic model is a decision maker making a decision by choosing one alternative, informed by analysis, which results in one outcome out of the many possible.

Though the word "prescriptive" is used to describe this type of decision analysis, the purpose of analytic support is not to tell the decision maker what to do. There are always going to be gaps between the model and reality because the real world contains a lot of complexity. These gaps mean that the analytic model will never perfectly describe the decision maker's situation, but imperfect analysis is often still useful. Hence, the purpose of prescriptive decision analysis is to provide analysis so the decision maker can make a more informed decision, but the choice still belongs to her. Besides, decision makers don't like being told what to do anyway!

By informed decision, I mean one that uses the information at hand systematically and effectively – one that is arrived at through a good decision process. It's important to draw a distinction between good decision processes and good outcomes, because a good decision process does not guarantee a good outcome (see Figure 9.1). A decision maker can make a good decision and have his efforts rewarded with a good outcome, or he can have bad luck and get a bad outcome. Similarly, a decision maker can make a poor decision (ignoring relevant information) and get a good outcome by being lucky, or he can get what he deserves in a bad outcome. It's much more fair to judge decision makers on their decision processes than on their outcomes because the decision process is the only aspect under their control. However, in real life, outcomes are usually easier to observe and are frequently used as a basis for appraisal.

In Section 1.3.1, I discussed how to identify opportunities for analysis. Identifying and framing a decision to inform with analysis is a similar skill with correspondingly strong impacts. If the decision has not properly framed, the analysts' effort may be wasted even if the analytic product is of high quality. An appropriately framed decision leaves room for creative thinking, is scoped to match the decision maker's authority and provide him with useful information, and has hooks for the analysts to model. The remainder of this chapter pertains to the action the analysts take after the decision has been chosen.

In Section 9.2, I will describe the decision environment, which is the situation that a decision takes place in. The Sections that follow (9.3 through 9.5) concentrate on a classic and versatile decision analysis technique called multi-attribute utility functions. In Section 9.6, I will discuss decision trees, which

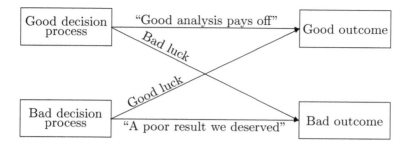

FIGURE 9.1
The relationships between decision-making process, outcomes, and luck. The visualization of this concept is adapted from work by Barry Ezell.

are a visual and intuitive method appropriate for modeling some decisions. Next, Section 9.7 is a brief introduction to game theory, a technique useful for modeling multiple decision makers whose decisions are interdependent. Finally, the chapter concludes with an introduction to negotiations, situations in two or more which decision makers may work together for their mutual benefit.

9.2 Understanding the Decision Environment

As analysts, when informing a decision it can be tempting to jump immediately into mathematics and model construction, but there are a few tasks that must be completed before making this leap. Most importantly, analysts must study and understand the environment in which the decision is taking place as well as the objectives of the decision maker. These may seem obvious at the start of the project, but there are often subtle aspects of the environment and the objectives that should influence the analysis. It's much harder to adjust and incorporate these aspects after the analysis has begun, so putting in effort at the front end saves time and effort in the long run.

The <u>decision environment</u> encompasses the people, processes, information, and constraints that influence or are affected by the decision and its accompanying analysis. I will describe some major aspects of the decision environment, but the concept is extremely broad in that it includes everything that touches or is touched by a decision. These environmental factors should influence the decision support provided by the analyst.

The central figure in the decision environment is the decision maker – he or she is the ultimate judge of whether the analysis is satisfactory and complete. For this reason, the decision maker is sometimes called the "customer" of the analysis. In the simplest situation, a decision maker is a single person who meets directly with the analysts, consumes the analysis, and has the authority to make the decision. In reality, this idealized situation rarely occurs; either the decision maker is actually a committee, the decision must be vetted by some other person or group who the analysts aren't in contact with, or the decision maker isn't actually available to meet with the analysts. These aspects of the decision environment should be taken into account when designing the type of analytical support provided and, especially, the manner in which the support is presented (see Chapter 11 for a thorough discussion of analytic communication).

Aside from the decision maker, there are many other people that are part of the decision environment. These other people are collectively referred to as <u>stakeholders</u> – a broad term that includes the people and organizations who will be affected by the chosen alternative or can influence the outcome in some other way, as well as those who contribute to carrying out the analysis. This makes the members of the analytic team stakeholders themselves. In most cases, stakeholders care about the decision and are interested in the support the analysts provide to the decision maker. If possible, the analysts should involve the stakeholders in the decision by describing the analytic support and the process by which the decision will be made. The decision maker often also recognizes the importance of the stakeholders, and may want to include them in the decision-making process using briefings, town halls, dialogues, issue papers, and other outreach strategies. If stakeholders are excluded from the decision-making process, there is a larger chance of their criticizing the decision and the analysis if the decision maker chooses an alternative they object to or just don't understand.[2] Stakeholders who have been included in the decision process are to some degree less likely to object after the fact, but more importantly they often bring a different perspective and new information to the process. This tends to make the analysis more complete.

Sometimes, a decision takes place within an existing process, such as an annual budget cycle. These processes usually have an established rhythm for making decisions that the analysis must therefore respect. The various aspects of the process, such as the intermediate steps, required permissions and checkpoints, and the history of how similar decisions have been resolved should also be included as components of the decision environment.

The availability and quality of data is a part of the decision environment that strongly influences the analytical support that can be provided. Some decisions take place in data-rich environments, where there are observations that can be used to calculate all of the inputs into whatever model is built. Other decisions must be made in data-poor environments, where subject matter expertise and rough estimations are the only inputs available to inform

---

[2]This may happen even if you've included them. Don't take it personally.

the decision. Surveying to determine what data are available, including qualitative and quantitative data and subject matter expertise, is an important step before deciding on a plan for analytic support. In short, it is not helpful to design an analysis that requires the analysts to gather unobtainable data, so discovering what data are out there is helpful.

Certain pieces of information are critical features of the decision environment because they directly impact the analysis. One example of this type of information is a <u>constraint</u>, which describes either rules that the viable alternatives must satisfy or that the analysis itself must obey. Constraints are often imposed by the decision maker, but can also originate from law, organizational policy, the budget, and other sources. A typical constraint for business decisions is a ceiling on the total cost of the chosen alternative.

Additional constraints on the analytic support include the time frame in which the decision must be made (and, therefore, the analysis completed) and the resources available to conduct the analysis. Analytic support delivered to the customer after the decision has been made, violating the time frame constraint, is typically not useful. Executing an analysis also often requires staff with specialized skills, such as data gathering, information management, elicitation, and statistical analysis, which introduces its own constraint. The analysis must be able to be executed with the resources at hand, including staffing, data, and computational hardware and software. In Chapter 10, I will discuss the use of project management techniques for planning an analytic project so that it can be delivered on time – these techniques are very useful for explaining to the decision maker the tradeoffs between the information the analysis provides and the demands on the analysts.

After the decision environment has been fully studied, the analysts' understanding can be formalized in a written document. The process of writing this document brings the analytic team to a common understanding and can be referred to later on when questions inevitably arise. It can also be helpful to present the written document to the decision maker to confirm that he or she agrees with its findings and to demonstrate progress.

This has all been very abstract, so let's make things more concrete by describing an example decision that I will refer to through the next few sections. Imagine that I am going to purchase a new car. Before starting to model the decision, I will survey the decision environment. I'm the decision maker as well as the analyst, and there aren't any other stakeholders.[3] I'm going to use the car primarily to commute to work, but also for occasional trips out of town. This will be my only car and I will be the only person driving it. I have a budgetary constraint and also a constraint on when I must make my decision; I must have the car in my possession by the time my new job starts in two weeks. That's all the time I have to gather information, complete my analysis, and close a deal. All of these pieces of information constitute the

---

[3]There's an argument to be made that the companies competing to sell me the car are also stakeholders, and I'm sure they would be happy to supply me with information regarding their products, but they don't have any decision authority.

decision environment, but no analytic modeling should proceed until I also explore my objectives.

9.3   Determining and Structuring Objectives

Decisions are complex and in many cases the decision maker is unsure which alternative he prefers – that's why analytic support is helpful. Part of the complexity arises because decision makers often have multiple objectives they would like to achieve with a single decision. For example, when I buy a car I have several objectives: I want the car to be safe, look stylish, and fit within my budget, amongst other criteria. Unfortunately, the car that looks the most stylish is rarely the safest, and on the slim chance I find both of those characteristics in a single car, it probably won't fit within my budget.

Since one alternative rarely is superior on all the objectives the decision maker cares about, I need a basis on which to evaluate the alternatives and determine which one best satisfies the decision maker. That choice of basis is really important – it has a strong influence on which alternative will be preferred. For example, consider your objectives when choosing an apartment. The apartment's price, size, neighborhood, features, and proximity to your workplace all influence your decision. Now imagine that you carried out two separate analyses: one that included all the objectives and another that neglected to include proximity to your workplace. Those two analyses could potentially return two different recommendations for your choice of apartment. However, the second analysis would be missing something important, something the decision maker cares about. That's why analysts need to study the decision maker's objectives and build a complete basis for evaluating the alternatives.

In this section and the two that follow, I will introduce the decision support techniques known as value-focused thinking (Keeney 1993) and multi-attribute utility functions. Value-focused thinking is used to clarify and structure the objectives of the decision maker; these objectives are the basis on which the alternatives are evaluated. Multi-attribute utility functions then provide a mathematical framework with which to evaluate the alternatives against the objectives. When used together, the two techniques provide a structured way to determine which alternative the decision maker prefers.

Let's start by being a bit more deliberate about what I mean by an objective. An objective is something the decision maker wants to achieve by making a decision or, alternately, something he wants to prevent. In decision analysis, properly formed objectives have two components: a direction and a noun. Some examples of objectives are "minimize cost," "maximize happiness," and "minimize commuting time." The objectives are connected to the decision and its environment, meaning that if I'm making two different decisions I'm likely going to have different sets of objectives for each of them.

How are the objectives for a decision generated? In short, in value-focused thinking they originate from the values of the decision maker and the other stakeholders. Each person should write out, in short phrases, what they hope to achieve with the decision. Think about what would be the best possible outcome from the decision – what qualities make it the best? Think about what would be the worst possible outcome – what qualities make it the worst? Keeney (1993) describes additional thought processes that can be used to creatively generate objectives. After going through the objective identification exercises independently, the stakeholders should be brought together to engage in a discussion regarding the objectives they have written down. This discussion will usually result in the identification of new objectives that no stakeholder identified independently.

Even though the various stakeholders may have antagonistic relationships, surprisingly often they can agree on their objectives. For example, both a power company and an environmentalist group would like to maximize power generated and minimize environmental impact. Conflict usually results not from the objectives themselves, but from the differing tradeoffs the parties place between the objectives. Tradeoffs will be discussed in more detail in Section 9.5.

After the objectives have been generated, the next step is to organize them. Collect all the objectives into a master list. The analysts should start by combining similar objectives that are essentially synonyms. For example, in most situations "maximize savings" can be absorbed into "minimize costs." Similarly, "minimize travel time" and "minimize commuting time" can be combined for a decision about where to live.

When that is complete, the next part of organizing the objectives is to categorize them. Objectives can typically be categorized into two types: means objectives and fundamental objectives.[4] Fundamental objectives are the bottom-line reasons to make a decision. When a person asks himself why he cares about a fundamental objective, the answer is something along the lines of "I just do." On the other hand, means objectives are important because they lead to fundamental objectives – they are the means to an end. When a person asks himself why he cares about a means objective, the answer is that it is connected to some other objective, which can be fundamental or another means objective. Eventually, if you follow it far enough, a chain of means objectives will lead to one or more fundamental objectives. Whether an objective is categorized as means or fundamental depends on the nature of the decision; in other words, the same objective could be fundamental for one decision and means for a different decision.

The reason to differentiate between means and fundamental objectives is

---

[4]There are some exceptions to this categorization that I will not discuss in this book, such as strategic objectives.

that the fundamental objectives, when properly organized, are the basis for evaluating the alternatives. The means objectives, in the vast majority of cases, should not be used for this purpose.[5] If the fulfillment of means objectives were to be evaluated along with the fundamental objectives, some aspects of an outcome would be double counted. For example, for my car purchase example imagine I had a fundamental objective to minimize the total cost of ownership. If I also included an objective to minimize the amount I spent on gas, I would be double counting the satisfaction I receive from spending less money through owning an efficient car as well as my dissatisfaction with owning an inefficient car. That double counting would skew the results of my analysis. So, taken as a whole, the set of fundamental objectives must be mutually exclusive and collectively exhaustive. If the set is not mutually exclusive, aspects of an alternative will be double counted. If the set is not collectively exhaustive, aspects of the alternatives that are important to the decision maker will be omitted from the evaluation.

Once the analysts have separated out the fundamental objectives, the next step is to organize them into a hierarchy. The top of the hierarchy is the overall fundamental objective; all the other fundamental objectives are its subcomponents. In the car purchasing example, my overall fundamental objective is to maximize my satisfaction with the new car. The analysts may need to create the overall fundamental objective; it is often too general for one of the stakeholders to include it in their lists. The hierarchy extends downward in treelike form from the overall fundamental objective, with each level satisfying mutual exclusivity and collective exhaustion. The fundamental objectives hierarchy ends when it reaches a level where the alternatives can be evaluated to find outcomes; this evaluation is discussed in detail in Section 9.4.

Building an appropriate fundamental objectives hierarchy is difficult and takes serious time and thought. It's hard in a different way than some of the more mathematical parts of this book are hard – crafting the objectives so they succinctly convey stakeholder concerns exercises a different part of your analytical brain. The time you spend crafting the objectives hierarchy is well-rewarded later in the process. Assembling a complete and well-structured objective hierarchy lays a good foundation for the decision model and makes the analysis and communication much easier later on.

Means objectives are organized into networks, rather than hierarchies. The means objective network connects to the fundamental objective hierarchy in that the fulfillment of means objectives results in the fulfillment of fundamental objectives. There's no analytical requirement to build the means objectives network, but it can be helpful when stakeholders who participated in identifying objectives ask the analysts where some of their suggestions went.

Consider how the scope of the decision influences the fundamental objec-

---

[5]The exception to this rule occurs when the fundamental objectives cannot be measured, even with a constructed attribute. In that case, there is no choice but to measure a means objective.

tives, as illustrated by my example decision of buying a new car. Suppose that instead of framing the decision as which new car to purchase, I instead framed it as how best to commute to work. This opens some new possibilities, such as riding my bike to work. In this case, perhaps I would have included "maximize exercise" and "minimize commuting time" as objectives. Then I could have considered the tradeoffs between buying a car and riding a bike. This example illustrates that the analysts need to pay careful attention to how the decision is framed because it influences the fundamental objectives, which in turn influence the preferred alternative.

After creating the fundamental objectives hierarchy, the analysts should identify or build scales with which to evaluate the alternatives against the objectives. These scales are called attributes – I've already discussed the three types of attributes (natural, constructed, and proxy) in Section 2.4. It's important to assign an attribute for each fundamental objective on the level of the hierarchy where evaluation will take place to fully capture all of the reasons the decision maker cares about the decision.

Some objectives take little effort to assign attributes. For example, cost often has a simple natural attribute of dollars (or whatever currency is most appropriate). Assigning attributes to some other objectives requires more effort from the analysts, such as when a scale needs to be constructed. Make sure the extremes of the scales are wide enough to capture the variety of outcomes possible. Keeney and Gregory (2005) specified five desirable properties for attributes; they should be unambiguous, comprehensive, direct, operational, and understandable. The operational aspect of an attribute is key – the data to populate that attribute must be obtainable by the analysts.

In Figure 9.2, I've written out a fundamental objectives hierarchy for the example of buying a new car. Notice how the fundamental objectives flow downward in the hierarchy from the overall fundamental objective. No attribute is necessary for the overall fundamental objective because the reasons the decision maker cares about the decision are captured by the mutually exclusive and collectively exhaustive set of fundamental objectives below. Let's walk through each of these fundamental objectives and identify some potential attributes:

- Minimizing the total cost of ownership is my overarching monetary objective, which I will measure with an attribute of dollars spent. To calculate this attribute I'll make a simple assumption regarding how long I think I will keep the car – let's say 10 years. The cost estimate then proceeds by combining the cost of gasoline the car will use, the car's resale value after 10 years, and the estimated cost of 10 years of insurance, repairs, and tune-ups, etc.

- Minimizing impact on the environment is important to me. Though cars have more environmental impact than is caused just by their carbon emissions, I will use estimated miles per gallon as my attribute for this objective. The monetary aspects of using less gas are included in the minimize cost objective, so here I would consider only the environmental aspects of my satisfaction.

- Maximizing style is my all-encompassing objective for how much I like the look of the car. It's not necessary to spend a lot of time writing out exactly what I do and don't like; for this decision it's enough just to use a constructed scale to build a simple ordinal rating system populated with my personal preferences.

- Maximizing driving enjoyment is a complex fundamental objective that includes the handling, power, and features of the car. Again I will construct a simple ordinal ranking system populated by my personal preferences, ideally determined after taking each alternative for a test drive.

- Maximizing reliability is included as a fundamental objective not because repairs are expensive (that's included in the minimize cost objective) but because it's a hassle for my car to be in the repair shop when I need to use it. As an attribute I could use either an estimate of the number of major repairs needed over the 10 years I'm going to keep the car or the reliability score it received from a car magazine.

- Maximizing safety is included as a fundamental objective for obvious reasons. As an attribute, I could use either the National Highway Traffic Safety Administration's safety ratings (an ordinal star system) or a personally designed constructed ordinal scale built on the presence of the safety features I am interested in.

- Maximizing cargo room is important to me because larger capacity makes the car more versatile for my trips out of town. A good attribute is cubic feet of cargo space in the trunk.

When I was completing this exercise, I also identified some objectives that I later classified as means objectives. These included "maximize the number of airbags," "maximize engine power," and "maximize number of miles between tune-ups." Each of these means objectives had one or more fundamental objectives they were connected to. If I had included them as fundamental objectives, I would have been double counting some aspects of the decision.

In most cases, the process of determining and structuring objectives occurs simultaneously with developing an understanding of the decision environment. This happens because both of these processes advance through discussion with the decision maker and interested stakeholders. In other words, as the analysts observe the decision environment they also get a better sense of the various parties' objectives.

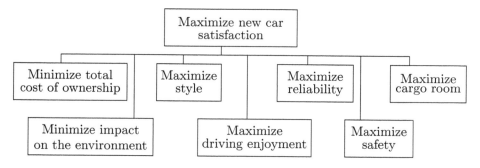

FIGURE 9.2
The fundamental objectives hierarchy for the car purchase example. The overall fundamental objective is located at the top of the hierarchy. The other
fundamental objectives are split into two vertical levels solely so they will fit
on the page.

9.4   Identifying Alternatives

After understanding the decision environment and structuring the decision
maker's objectives, the analysts should identify a set of alternatives for evaluation. Examples of sets of alternatives include:

- Various neighborhoods to live in for someone who has just moved to a new
  city,

- Different resource allocations for a manager budgeting for the next year, and

- Different experiments for a lab director to pursue.

In Section 9.5, I will explain how to use a multi-attribute utility function to
determine which alternative is preferred, but before that step a broad set of
alternatives must be identified and evaluated to produce the outcomes. This
subsection is based on the discussions in Keeney (1993) and Hammond et al.
(1999).

Why should analysts expend extra effort to identify a broad set of alternatives? Can't we just identify a handful and move on to evaluating which is
preferred? Absolutely not – as Hammond et al. (1999) says, "you can never
choose an alternative you haven't considered" and "no matter how many alternatives you have, your chosen alternative can be no better than the best
of the lot." If the analysts fail to identify a strong alternative that would otherwise be available, then the decision maker may miss out on potential gains.
For example, a restaurant owner looking to open a new location may focus on
renting a space, never exploring the possibilities of opening a mobile location

or buying a space. Creative alternatives like these are not always preferred, but sometimes something surprisingly attractive pops up.

One of the most common mistakes when performing prescriptive decision analysis is jumping too quickly to identifying alternatives; Keeney calls this mistake "alternative-focused thinking." The reason this is a mistake is that it usually results in only obvious alternatives being identified. It leads people to focus on the status quo and incremental adjustments thereof (Keeney 1993).

Only after understanding the decision environment and building an objectives hierarchy should the analysts begin identifying a set of alternatives. Though the obvious alternatives should also be captured, creativity is the key to this process – often the alternative that will best satisfy the fundamental objectives is not obvious and takes some inspiration to generate. It is much easier to find creative alternatives by thinking about how to achieve a concrete set of objectives rather than by thinking about the situation at hand in a free form way.

Even if the analysts try to delay identifying alternatives until after the objectives hierarchy is complete, a few different alternatives are likely to have surfaced in the course of the analysts' discussions of the decision environment and objectives with the decision maker and other stakeholders. These are usually the most obvious alternatives, but that doesn't mean they are not valuable. Make sure you record them when they appear, because the person who suggested an alternative will be looking to make sure that it has been included in the set and is evaluated against the other alternatives.

The decision maker usually has a few alternatives he would like included in the analysis. Most decision makers are not shy about telling the analysts the alternatives they would like included, but the analysts should make sure the decision maker is given the opportunity to contribute regardless.

Once the obvious alternatives have been identified, how can the analysts use the fundamental objectives hierarchy to identify creative alternatives? Keeney (1993) describes a simple tactic – proceed through each of the fundamental objectives and identify an alternative that achieves that objective to the utmost, ignoring the alternative's merits on the other objectives. Next, identify alternatives that do the same for subsets of the objectives, starting with two objectives at a time and moving to larger groupings. This process will generally identify a few alternatives that were not apparent at the start of the process.

Another tactic for identifying alternatives is to directly gather them from minor stakeholders and people not involved in the decision. People with an outside perspective are especially valuable for identifying creative alternatives. Send out email requests, hold brainstorming sessions, and call colleagues. Ask them to describe some alternatives they might consider if they were placed in the same situation as the decision maker.

Keep in mind that some viable alternatives may not conclusively resolve the situation. For example, in many decisions a viable alternative is simply to wait longer without committing to one path or another and set a time to revisit the decision. Another viable alternative may be to gather more information before committing. Though not earth shattering, these can be reasonable choices and should be evaluated in the same way as the other alternatives. These types of alternatives typically have costs associated with them, and some of the other alternatives may no longer be available when the decision is revisited.

Incidentally, the identification of alternatives provides an easy way to build trust and credibility – if an especially creative alternative is suggested by someone who is not an authority figure, it can be helpful to associate that alternative with the person who suggested it, such as naming it "Bob's strategy." Giving credit for creative ideas builds good will. On the other hand, if the idea was suggested by someone in a position of authority, associating the alternative with the authority figure can give that alternative a sense of inevitability; this is something to avoid.

After the analysts have constructed the set of alternatives, they should present them to the decision maker to verify that they cover the full spectrum of what he is considering. This discussion may prompt the decision maker to create new alternatives that he would also like included in the analysis.

Let's generate some creative alternatives for the example of purchasing a car. The fundamental objectives do suggest some creative alternatives that I hadn't previously considered, such as biking to work (to minimize impact on the environment), leasing a car or buying a used car (to minimize total cost of ownership), and buying a motorcycle (to maximize driving enjoyment). Let's assume that, though interesting, I'm not all that attracted to any of these alternatives. Then, I would go to a few different car dealers and start doing research on different cars. I could look at different cars that maximize each fundamental objective, but I can use my judgment in cases where I don't think this will be helpful. For example, to maximize style I'd probably want to look at a sports car that is well out of my price range. There's no need to waste time on researching those alternatives (though test driving them could be fun).

After the set of alternatives has been finalized the alternatives should be evaluated against each of the attributes from the previous section – these evaluations are the projected outcomes of the decision. If there is uncertainty in any of the evaluations, regardless of its origin, that uncertainty should be communicated to the decision maker.

Table 9.1 illustrates the evaluation of three hypothetical cars, the Sporta, Cheapo, and Leafy, against my fundamental objectives and their associated attributes. Each of the three car alternatives has more than one objective on which it is superior. The Sporta is the best choice for style and driving enjoyment, the Cheapo is the best for cost, reliability, and cargo space, and the Leafy is the best for environmental impact and safety. At this point, it is not clear which alternative I prefer.

Frequently, the decision maker will be satisfied with an analysis that identifies alternatives and proceeds no further. Sometimes, the decision maker

TABLE 9.1
Three hypothetical car alternatives evaluated with attributes for the buying-a-car example.

Fundamental objective	Attribute	Car name		
		Sporta	Cheapo	Leafy
Minimize total cost of ownership	Dollars	$30,000	$15,000	$20,000
Minimize impact on the environment	Miles per gallon	15	25	45
Maximize style	Ordinal style scale	3	1	2
Maximize driving enjoyment	Ordinal driving scale	3	2	1
Maximize reliability	Reliability score	8	10	9
Maximize safety	Safety score	9	9	10
Maximize cargo room	Cubic feet	8	15	12

may even be content solely with the construction of an objectives hierarchy. That's fine – many management consultants provide exactly that service, and if the decision maker is satisfied it's a job well done. That said, many of the quantitative techniques in prescriptive decision analysis start from this point in the process, after the fundamental objectives have been assigned attributes and alternatives have been identified and then evaluated with those attributes. The next section discusses a quantitative technique for using these evaluations to determine which alternative is preferred.

9.5   Utility Functions

It seems like it should be straightforward to determine which alternative out of a set is preferred. For example, if you're choosing a job, the preferred alternative is the one that has the highest salary, right? Of course not; in real life, most decisions have multiple objectives, and since it is rare for an alternative to be superior on all the fundamental objectives, which alternative is preferred depends on the tradeoffs between those objectives. Additionally, some alternatives have uncertainty with regard to the resulting outcome. Building a utility function is one way to sort out the decision maker's preferences given all these complications.

In simple terms, a utility function is a ratio scale that quantifies the relationship between the decision maker's preferences and the available alternatives. You can think of the utility of an alternative as a measure of the decision maker's satisfaction or happiness. Traditionally a utility function is unitless and scaled so the worst possible outcome has a score of 0 and the best has a score of 100. Since a utility function is a ratio scale, outcomes that score 100 are twice as preferred as those that score 50, for example.

Why is it important that a utility function is a ratio scale? It's clear that for some decisions, using a ratio scale is excess information. For example, when I'm at a restaurant and I'm choosing which meal to order, I don't need to know that I place twice as much value on the hamburger as on the pizza. It's enough to simply know that I will enjoy the hamburger more than the pizza – then I can make my order.

One reason analysts use ratio-scale utility functions is that they become particularly useful when making decisions that have uncertainty in the outcomes. Let's illustrate this by beginning a mathematical discussion of utility functions – consider a decision maker concerned only with fulfilling a single fundamental objective. Perhaps the easiest fundamental objective to understand is a person who plays a game of chance trying to maximize his or her winnings.

Imagine you were presented with a simple game: flip a coin, and if it lands on heads you win \$2, and if it lands on tails you win nothing. How much would you pay to play this game? Well, on average, every time you play you win \$1, because half the time you win \$2 and the other half of the time you win nothing. This average outcome of the game is called the game's expected value.

If you can play the game for less than the expected value on average you'll be earning money, but if the cost to enter is more than the expected value on average you'll lose money. If the cost to play is exactly \$1, on average you'll break even. It won't work out this way for every flip of the coin, but in the long run odds are this is how the game will turn out. That's the general structure of games of chance.

Let's change the format of the game a bit. Now, pretend the person running the game offers you a choice: take a guaranteed payoff of \$0.50 or flip the coin for the same \$2 and \$0 payoffs as in the previous game. Let's assume that you value \$2 four times as much as \$0.50 and analyze the outcomes in terms of a utility function U. There are three possible outcomes: $U(\$2) = 100$, $U(\$0.50) = 25$, and $U(\$0) = 0$. Then, the expected utility (defined similarly to expected value) of choosing to flip the coin is 50, which is larger than the expected utility of the guaranteed payoff. Therefore, to maximize your utility, you should choose to flip the coin.

The key to the previous example was the assumption that the ratios of the utilities matched the ratios of the payoff amounts, i.e., that the utility function mapped linearly from payoffs to utility. In the more general case, the ratio of the utilities is subjective in that it is dependent on the decision maker's preferences and sometimes those preferences can be counterintuitive. For example, let's say I have a choice between a guaranteed $100 million and a coin flip with prizes of $300 million and $0. The coin flip has an expected value of $150 million, so I'm an idiot if I choose the sure $100 million, right?

Absolutely not. Even though $300 million is three times as much money as $100 million, the two sums have almost the same amount of utility to me because they are both way more money than I have right now. This is the essence of what makes a utility function subjective – if you asked a billionaire to make this decision, he might well choose the coin flip instead of the sure payoff because the sure payoff of $100 million isn't life-changing money to him, and his reasoning would be perfectly logical. Even though we'd be choosing different alternatives, we'd both be making the correct choice given our personal utilities.

Technically, my utility function in this situation is known as risk averse, which essentially means that I'd rather have the sure payoff even if I'm making a sacrifice compared to the expected value of the coin flip. A decision maker who chooses based solely on the expected values, like the billionaire in the example did, is said to be risk neutral.

Are there situations where the decision maker is risk seeking – i.e., where he or she prefers the coin flip even if its expected value is less than the guaranteed payoff? Absolutely – in its most extreme form, this is the thought process many people use when entering their state lottery. The $1 entry fee buys a one in 50 million chance to win $10 million (for example). Based on expected values this lottery is clearly not worth it, but the small entry fee makes such a small difference in a person's utility that many people consider it worthwhile for even a small chance at life-changing money. Risk-seeking behavior occurs in many cases where the winning payoff allows the decision maker to do something that the guaranteed payoff doesn't get them close to achieving.

The three different attitudes toward risk are depicted in Figure 9.3. This figure is intended as a schematic representation of their utility functions and should be taken as an example only. Notice how for an intermediate payoff, the risk-averse decision maker values it relatively more than the risk-neutral and risk-seeking decision makers. Similarly, see how the risk-seeking decision maker's utility really starts to increase only toward the high end of the payoff range and the risk-averse decision maker's utility stays relatively flat over the higher end of the payoff range.

Utility functions aren't limited to continuous quantities like dollars; it's straightforward to build utility functions for discretized payoffs as well. For example, in my car purchasing decision I ranked the three car options by

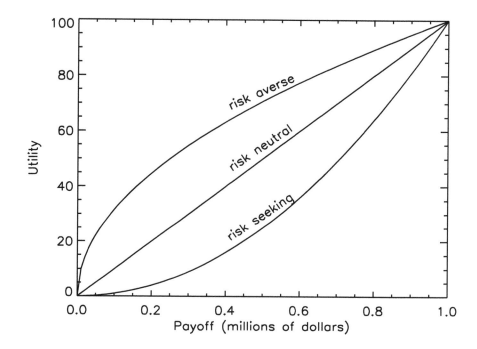

FIGURE 9.3
A schematic representation of different risk preferences.

driving enjoyment; the Sporta got three points, the Cheapo got two, and the
Leafy just one. To construct a utility function in this situation, assign the
extremes first; the Sporta scores 100 and the Leafy 0. Now I can assign a
utility to the driving enjoyment of the Cheapo based on my own personal
preferences. From my test drive, I found that the Cheapo and the Leafy drive
similarly, with my preference leaning slightly towards the Cheapo, while the
Sporta was better than both by leaps and bounds. So, I'll assign a utility of
20 to the Cheapo's driving enjoyment; that score feels about right to me.

   The bottom line takeaway of the discussion so far is that the utility function
depends on the preferences of the decision maker. The same outcomes could
be described to different decision makers and their relative preferences would
be totally different. You have to talk to the decision maker in order to figure
out what their preferences are.

   I've really just scratched the surface of some of the counterintuitive results
related to utility functions. Part of what makes this field so interesting is that
a lot of it is experimentally testable – this starts to get into the subject of
behavioral decision analysis. For a more detailed discussion of these issues, see
Chapter 3 of Raiffa et al. (2002).

It's relatively straightforward to build a utility function for decisions that involve a single fundamental objective. However, as we discussed in Section 9.3, most of the important decisions we make involve multiple fundamental objectives. This requires you to build a multi-attribute utility function, a model that includes all of the fundamental objectives, each measured with its own attribute, and combines them into an overall utility score.

In this more general case, the utility functions associated with each attribute are called the single-attribute utility functions, u_i. These u_i are developed in the same way as the utility functions described above – by talking to the decision maker. The single-attribute utility functions are also ratio-scale quantities that are scaled from 0 to 100. Even this simple structure requires making an assumption – that the fundamental objectives contribute independently to the overall utility.

The simplest form for a multi-attribute utility function is linearly additive:[6]

$$U(x) = \sum_{i=0}^{n-1} w_i u_i(x), \qquad (9.1)$$

where x is the alternative being evaluated, $u_i(x)$ is the single-attribute utility for the outcome of alternative x, and n is the number of fundamental objectives each of which has its associated swing weight, w_i. The multi-attribute utility function is just a weighted sum over the single-attribute utilities.

So what is a swing weight? Most importantly, it is *not* an importance weight, which is what most people think of when they consider weighting objectives. An importance weight is the answer to the question "how many more times important is objective A versus objective B?" Importance weights are not a good way to build a utility function, as the following example will demonstrate. Consider you are deciding between two jobs and you have two fundamental objectives: maximize your salary and live in the warmest city. The first alternative pays $50,000 per year and is located in Los Angeles, and the second alternative pays $50,100 per year and is located in Anchorage. If you had said that salary was your most important objective, this methodology might tell you that the Anchorage alternative is preferred. This seems wrong, because the difference in salary is quite small and the difference in the warmth

---

[6]There are good theoretical reasons for a properly crafted set of fundamental objectives to mesh with a linearly additive multi-attribute utility function (Stewart 1996). However, there are certainly occasions when other forms of a utility function are a better fit. In particular, this occurs when the attributes of the fundamental objectives interact to affect the decision maker's overall utility. For example, consider the decision of choosing a job which included two attributes: the number of hours of work and the subject matter of the work. If the subject matter was something you found boring, you might prefer to work the standard 40 hours a week. However, if you found the subject matter interesting, you may prefer to work 60 hours per week. The utility you derive from one of the attributes depends on the value of the other one; therefore, they are not independent and a linearly additive utility function would not be appropriate. These situations can also be modeled with other utility function forms (Keeney 1993).

of the locations is enormous. Importance weights do not take these differences into account, but swing weights do.

Swing weights are mathematical descriptions of the relative importance of changing from one extreme to another for each attribute; they are the decision maker's articulation of his tradeoffs across the objectives. Swing weights are determined from the decision maker's values. The decision maker must determine the relative utility of changing from the low to the high extreme of one attribute versus the same type of swing for another attribute. In other words, the decision maker could say that changing from the best to the worst total cost of ownership is worth the same as changing from the best to the worst reliability. Then the swing weights for these two objectives would be equal. Alternatively, he could say that the former is worth twice as much as the latter; then the swing weights would have a 2:1 ratio. After the relative values of all the swing weights have been determined, they are then normalized so that they sum to 1: $\sum_i w_i = 1$.[7]

Let's illustrate this concept with a final discussion of the car purchase example. First, examine how I've assigned the single-attribute utilities in Table 9.2 – in particular, look at the style and driving enjoyment objectives. By assigning a single-attribute utility of 80 to the leafy for style, I'm stating that in my view the style of the Leafy and the Sporta are much closer together than that of the Leafy and the Cheapo. In contrast, by assigning a single-attribute utility of 20 to the Cheapo for driving enjoyment, I'm stating that the Cheapo and the Leafy are much closer together than the Sporta and the Cheapo.

The swing weight I've placed on the safety objective is relatively low not because I don't think safety is important, but because I don't see much of a difference between safety scores of 9 and 10. Similarly, the swing weight I've placed on the cost objective is relatively high because there is a large difference in my value of the best and worst alternatives. This is similar to the earlier example of a job choice dependent on salary and city, with safety playing the role of salary and cost playing the role of city.

Using the swing weights I've assigned, the bottom row of Table 9.2 shows the calculated multi-attribute utilities for each of the three alternatives. The Cheapo is preferred slightly to the Leafy, which are both strongly preferred to the Sporta. If I were presenting this analysis to a decision maker (instead of making the decision myself) and the utilities and objectives had been built from his values, I would add some caveats to the finding of a slight preference, since the difference between the utilities of the Cheapo and the Leafy is less than 10% and the precision of the swing weights and the single-attribute utilities is not perfect. It could be useful to perform a sensitivity analysis on the swing weights where the values are adjusted to observe the effect on which alternative is preferred (see Clemen and Reilly (2001) for more information). As an analyst, I'd feel like I did my job well if the decision maker chose either

---

[7]For readers looking for more detail on how to determine swing weights, Clemen and Reilly (2001) includes a structured methodology.

TABLE 9.2
Three hypothetical car alternatives evaluated with attributes for the buying-a-car example. Single-attribute utilities (in parentheses after the outcomes), swing weights, and multi-attribute utilities have been added to this table.

Fundamental objective	Swing weight	Sporta	Car name Cheapo	Leafy
Minimize total cost of ownership	0.3	$30,000 (0)	$15,000 (100)	$20,000 (66)
Minimize impact on the environment	0.2	15 (0)	25 (60)	45 (100)
Maximize style	0.1	3 (100)	1 (0)	2 (80)
Maximize driving enjoyment	0.1	3 (100)	2 (20)	1 (0)
Maximize reliability	0.2	8 (0)	10 (100)	9 (50)
Maximize safety	0.05	9 (0)	9 (0)	10 (100)
Maximize cargo room	0.05	8 (0)	15 (100)	12 (50)
Multi-attribute utility		20	69	65

the Cheapo or the Leafy, but I would feel like I made some mistakes if the decision maker saw this analysis and still chose the Sporta.

9.6   Decision Trees

Decision trees are structures used to model and visualize decisions. They are particularly useful because decision makers who are not trained analysts tend to find them relatively intuitive. Decision trees are often used in conjunction with the utility functions described in the previous section, though they are also applied along with many other ratio-scale quantities, such as monetary payoffs.

A decision tree is read from left to right and is constructed from nodes and

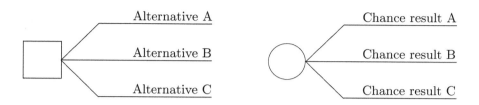

FIGURE 9.4
A decision node (left) and a chance node (right).

branches. The nodes represent the introduction of choice or chance into the model, while branches depict the different options that can result from each node. At the very end of the rightmost level of branches are outcomes. The set of branches that emerge from each node must be mutually exclusive and collectively exhaustive.

The visual iconography of decision trees is relatively standardized, with decision nodes represented by squares and chance nodes represented by circles. Examples of the two types of nodes are illustrated in Figure 9.4. On the left-hand side, I have drawn a decision node with three emerging alternative branches. At a decision node, the decision maker chooses which branch to go down. These branches must represent all of the alternatives available to the decision maker, and he or she must have to choose one and only one. On the right-hand side, I have drawn a chance node with three emerging chance branches. At a chance node, the branch that occurs is random and some branches can be more probable than others. These three branches must represent all of the possible outcomes from the chance event, and one and only one of them must occur. All branches, if they don't lead to another node, end in an outcome.

The node and branch structure of a decision tree is particularly appropriate for a discretized set of outcomes. Representing continuous outcomes is a struggle in terms of visualization because you can't draw an infinite number of branches. The mathematics of a continuous node can also be more complicated to solve.

The power of decision trees comes from the ability to chain together decision and chance nodes to model complex decisions. If you've built a utility function, or if the decision you're informing has a single ratio-scale attribute, you can use a decision tree to determine which alternatives lead to the highest expected utility. Decision trees are particularly good at representing sequential decisions made by a single decision maker with chance events occurring in between. Decision trees are often used to identify the alternative that will provide the largest expected utility (or expected value if a utility function is not used). This process is called solving the decision tree. Solve the tree from the right to the left, resolving the chance and decision nodes one at a time. For a chance node, the node is resolved by reducing it to its expected utility; for each branch, calculate the multiple of the relative probability of its occurrence and its utility. Sum these multiples for all the branches to get the expected utility. For a decision node, simply choose the branch with the highest expected utility – this represents the decision maker choosing the branch that he prefers.

Let's build a simple example to illustrate the process of solving a decision tree. Consider a person deciding whether or not to play a two-step game. It costs $1 to play the game. The first phase of the game is a coin flip. If the coin lands showing heads, the player loses their entry fee and if it lands showing tails, the player enters the second phase of the game. In the second phase of the game a playing card is drawn from a standard shuffled deck. If that card is a face card (an Ace, King, Queen, or Jack) the player wins $10. If any other card is drawn the player's entry fee is returned. Is it worth it to play this game?

The decision tree for this game is shown in Figure 9.5. To solve the decision tree, start with the rightmost node – in this case, the card draw chance node. Sixteen out of the fifty-two cards in a deck are face cards, which means the probability of drawing one is 4/13. It follows that the probability of not drawing a face card is 9/13. Therefore, the expected value of this node is $40/13. Moving leftward, in a fair coin flip there is a 50% chance of heads and a 50% chance of tails. This makes the expected value of the coin flip node ($20/13) − ($1/2) = $7/26. The decision node is then a choice between not playing, with an expected value of $0, and playing, with an expected value of $7/26. Since the expected value of playing is larger than not playing, the decision maker should choose to play (assuming a risk-neutral utility function).

For a more realistic example, consider the choice of whether or not to attend a professional baseball game. As Bull Durham said, in baseball "sometimes you win, sometimes you lose, and sometimes it rains."[8] For argument's sake, let's say the relative probabilities of these three outcomes for the home team are 50%, 45%, and 5%. Let's also say that you've studied the utility you place on the three outcomes, including the price of the ticket, your levels of enjoyment, etc., and scored them as 100, 20, and 0, respectively (arriving at the stadium only for the game to be rained out is very annoying). If you don't attend the game, let's say your alternative is to go to a movie instead, an activity with a utility of 50. The decision tree for this situation is shown in Figure 9.6. In this case, the decision to attend the baseball game has an expected utility of 59, higher than the movie.

---

[8]For those readers unfamiliar with baseball, in rainy weather the game is rescheduled to another date.

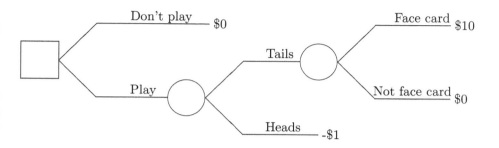

FIGURE 9.5
A decision tree for the coin flip and card draw game.

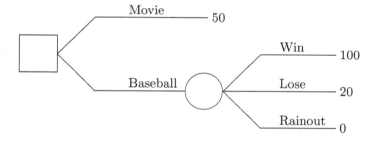

FIGURE 9.6
A decision tree for attending a baseball game.

9.7 Game Theory

Game theory is a branch of mathematics that models situations where two or more decision makers make choices that jointly influence the payoffs. Mathematicians have studied game theoretic problems for hundreds of years, but modern game theory really took off beginning with Von Neumann and Morgenstern (1944). That book spurred a lot of interest in the field, eventually leading to the application of game theory to problems in economics, military and political science, and many other subjects.

What sorts of decisions are amenable to support with game theoretic analyses? The primary indicator that a game theoretic methodology is an appropriate choice for prescriptive analytics is when the decision maker's optimal alternative depends on the choices of other people. Contrast this situation with decision trees, where there is typically one decision maker with some chance events thrown in. It is sometimes possible to model others' choices as probabilistic events (e.g., as a random node in a decision tree), but looking at all the decision makers' perspectives strategically can lead to new insights.

On the other hand, game theoretic methodologies can be less appropriate for some other decisions. Typical game theoretic analyses optimize decision making for a utility function (though not necessarily the same utility function for all the competitors). In theory, a decision maker's utility function can be created from his fundamental objectives hierarchy, but in practice this is often impossible because of unwillingness by one of the decision makers to quantity his or her tradeoffs (typically the decision maker the analysts are *not* advising) or just because of limited access to the decision maker the analysts are advising. These are practical impediments to using game theory rather than methodological ones, but they are problematic regardless.

The prisoner's dilemma is a classic problem that illustrates some key game theoretic concepts. The setup of the game is as follows – imagine that you and a friend have been accused of robbing a bank. The authorities have enough evidence to convict both of you of the lesser crime of possessing a gun without a permit, but not enough for the robbery itself. The authorities don't want a trial with an uncertain outcome, so they offer you both a deal. You each have been given two alternatives regarding the robbery charge: confess or plead not guilty. You and your friend must commit to your decisions without any knowledge of what the other has chosen. You are also not allowed to coordinate your actions by communicating prior to committing to your decisions. Each person's objective is to maximize his or her own utility, which means, in this case, minimizing their own prison time. Your friend's prison time does not influence your utility.

Once you and your friend have made your decisions, you both receive an outcome. If you and your friend both confess, you will be sentenced to four years in prison each for bank robbery and gun possession. If you and your friend both plead not guilty, you will be sentenced to two years in prison for gun possession. If one person confesses and the other does not, the one who confesses goes free and testifies against the other, who is sentenced to eight years in prison. Figure 9.7 presents these alternatives in tabular form. Each box represents a possible outcome of the game – which box actually occurs depends on the choices of the two players.

Should you confess or plead not guilty? Let's examine the decision from your perspective. If your friend confesses, it is better for you to confess as well, since you prefer four years of prison time to eight. If your friend pleads not guilty, it is also better for you to confess, since you prefer zero years of prison time to two. Thus, regardless of what your friend chooses, your outcomes are better if you confess! In game theoretic terminology, the confess alternative dominates the pleading not guilty alternative.

Your friend's decision-making payoffs are identical to yours, so it is also

Friend's Decision

		Confess	Plead not guilty
Your Decision	Confess	4, 4	0, 8
	Plead not guilty	8, 0	2, 2

FIGURE 9.7

Illustration of the payoffs in the prisoner's dilemma game. The payoffs list the number of years of jail time for you, followed by the number for your friend.

dominant for him to confess. Therefore, if both you and your friend follow this logic, both of you will confess and the payoffs will correspond to the top left corner of the table. This is an interesting result, since the "best" outcome for you and your friend would be for both of you to plead not guilty.

Note that the outcomes are represented here entirely in terms of years of prison time. This may not accurately reflect your utility – for example, if you'll be wracked with guilt for ratting out your friend if you confess and he pleads not guilty, your payoff isn't quite as simple as zero years in prison. This illustrates why capturing all of your objectives in the utility function is important.

As described here, the prisoner's dilemma is a single round game, but you can imagine versions where the players repeat the game many times in a row. This changes the strategy, because players can signal each other with their choices and attempt to settle into the "best" outcome. This kind of communication is known as collusion. In the repeated prisoner's dilemma game, there are also interesting effects that occur when the end of the game is approaching. For such an uncomplicated set of rules, there is a rich amount of detail that emerges.

The prisoner's dilemma is a simple situation that demonstrates some of the intricacies inherent in game theory. However, it is just one of the many situations that game theory can be used to model. For example, in the prisoner's dilemma, both players are aware of the other's alternatives and utilities. This situation is called perfect information, and it is often not reflective of real decisions, but decisions with imperfect information are generally more difficult to model. In the prisoner's dilemma, the decision makers choose their alternatives simultaneously; many decisions are made sequentially, and there are models available for these situations as well. The prisoner's dilemma is also an example of a symmetric game – where the decisions, payoffs, and utilities available to each player are identical. In reality, asymmetric games are more common.

If you are interested to learn more about game theory, my favorite introduction to the subject is in Chapter 4 of Raiffa et al. (2002).

9.8   Negotiations

A negotiation is a special type of decision in which two or more decision makers work jointly to come to a decision. The outcome each party receives depends on the decision made jointly by all. During the negotiation, the parties are allowed to communicate in whatever manner they see fit.

Negotiations are similar to game theoretic situations, with the important distinction that all parties (or some subgroup, such as three fifths of the Senators in the U.S. Senate, according to the rules of that group) must come to agreement on a chosen alternative. In game theory, in contrast, the parties typically make independent decisions and interdependent payoffs result.

Negotiations occur in many aspects of our lives, from financial transactions such as buying a house or a car to professional discussions such as accepting a new job. They also commonly occur in business situations, such as between purchasers and suppliers or when establishing a subcontracting relationship. In all of these contexts, negotiations can be hugely complex affairs involving tradeoffs across a multitude of objectives. As an analyst, you will sometimes need to perform research and analysis to inform an organization's negotiation strategy, or you may want to perform this kind of work yourself in support of your own negotiations. In this section, I will give you an introduction to the type of analysis that can be used to model and support negotiations. For a more in-depth discussion, Raiffa et al. (2002) has a rare combination of analytic rigor, strong pedagogy, and folksy wisdom. I've drawn the technical material in this subsection primarily from their work.

When most people think about a negotiation, they concentrate on maximizing their own utility at the expense of the other party. Posturing, escalating threats, and arguments are a frequent result of this kind of thinking. This part of a negotiation is called claiming individual value, but negotiations have a lot more to them than this zero-sum portion. During negotiations, the parties can also create joint value through exchanges that, when taken as a whole, leave them all better off. Identifying opportunities to create joint value can be difficult when behavior linked to claiming individual value is occurring simultaneously. Open and honest communication can help diffuse this tension, but such communication isn't always possible.

The topics that are resolved over the course of a negotiation, such as the share of the profit that each party will receive, are known as <u>issues</u>. Issues are not limited to continuous quantities, but can be discretized as well. In order to compare the possible negotiated agreements, it is necessary to develop a utility function for the party the analysts are supporting with analysis. One simple way to do this is to use issues and their associated attributes as replacements for the fundamental objectives and their associated attributes in Equation 9.1. The swing weights for the issues are found in similar fashion. Of course, more complex utility functions are possible for negotiations, just as they are for solitary decision makers.

To prepare for a negotiation, each party should study their available alternatives and identify their "best alternative to a negotiated agreement," or <u>BATNA</u>. As you might expect, the stronger your BATNA, the more leverage you have during a negotiation, because if the negotiated agreement doesn't benefit you as much as your BATNA you can walk away from the table.

As described in the first few sections of this chapter, identifying the BATNA can be made easier by developing an objectives hierarchy, attributes, and multi-attribute utility function for your decision. Though this analytical structure can be helpful in clarifying your thinking, the party you are negotiating with will likely have a different objectives hierarchy in mind. Furthermore, the dimensions over which the negotiation takes place are unlikely to be the same attributes identified for your fundamental objectives. For these reasons, the modeling of negotiations is not structured around objectives, but instead around issues, as described above.

Once the BATNA has been identified, it must be compared with the possible alternatives of the negotiated agreement. This comparison can be made using the concept of the <u>reservation value</u>, which is the relative utility of the BATNA on a scale where 0 is the worst possible negotiated agreement and 100 is the best. Stated more directly, the reservation value is the point at which it is worthwhile to accept a negotiated agreement. Converting the BATNA to a utility scale used to evaluate negotiated alternatives is difficult for the reasons discussed in the previous paragraph – the negotiation is probably not taking place over the same attributes on which the BATNA was evaluated. One way to make this simpler is to identify a negotiated agreement that you consider of equal value to the BATNA, then evaluate the utility of that agreement.

Let's see how these concepts work in action by examining a simple negotiation with two parties and a single ratio-scale issue: price. The negotiation could be over any type of sale, such as a car or a house. The buyer would benefit from any price below his reservation value (perhaps derived from purchasing an alternate car or using public transit), and the seller would benefit from any price above his (perhaps derived from selling to another party or keeping the car). Of course, the seller would prefer a negotiated price far above his reservation value, and the buyer would prefer a price far below his. The range of prices between the two reservation values is known as the zone of possible agreement (<u>ZOPA</u>), as these are prices from which both the buyer

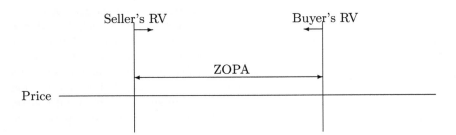

FIGURE 9.8

Diagram of the zone of possible agreement (ZOPA) and reservation values (RV) for a negotiation. The arrows underneath the RV lanes represent the range of prices each party will accept.

and seller would benefit. These concepts are depicted graphically in Figure 9.8. In practice, the parties can exchange offers until they settle on a price within the ZOPA.

What happens if the reservation value for the seller is higher than the buyer's reservation value? In this case, there is no ZOPA because there is no price from which both the parties would benefit and there is no possibility of agreeing to a deal, assuming the parties have accurately assessed their reservation values. This situation does frequently occur, especially with deals in which there is only one dimension in which to maneuver – not every negotiation opportunity has a viable solution, and the BATNA may very well be the best available option.

Compared to many negotiations, the single-issue example is an oversimplification because there are usually multiple issues on which to negotiate and trade value. Though these extra issues can make modeling the negotiation more difficult, they actually make agreeing to a deal easier because there is more flexibility and more possibilities to explore! In fact, one of the most interesting aspects of a negotiation is that the parties often have different utility functions and preferences. Then, each party can trade away things that they value less than the other decision makers and increase value for all parties. For example, one of the parties may be willing to concede ground on Issue A if it means they can gain ground on Issue B. Depending on the other party's preferences, they may value the change in B less and therefore also prefer this arrangement. In fact, when the parties trust each other, one productive technique is to exchange ranked lists of issue importance (taking into account the potential swing in that issue). In most cases, the two parties will have different preferences, making it possible to preferentially satisfy the issues on which each party places the most value.

Let's imagine you've negotiated with another party over a complicated set of issues and the two of you have identified an agreement that both of you would accept, i.e., it exceeds both of your reservation values. Does this mean the negotiations should end? No! Because negotiations are mechanisms for the parties to create joint value, there may exist some other negotiated agreement that provides more value to all parties. In other words, don't leave potential gains on the table. Keep talking even after you've found a potential agreement to see if you can improve on it.

9.9 Exercises

1. Describe one of the tougher life decisions you've had to make (choosing a college, a field of study, a job, a place to live, etc.). What were your fundamental objectives in making this decision?

2. Develop your own fundamental objectives hierarchy for buying a house. Evaluate three real houses as alternatives, assigning single-attribute utilities based on attributes you have chosen, assign swing weights, construct a multi-attribute utility function, and see which house the analysis says you prefer. Does the recommendation match what you expected before you performed the analysis?

3. Teenage boys have long played variants of a game called "chicken." Chicken is a two-player game in which the players drive cars directly at each other, then (sometimes) swerve at the last minute. Each player must make a choice between two alternatives: macho and chicken. The best outcome for a player is if he chooses macho and his opponent chooses chicken. Build a payoff matrix similar in style to Figure 9.7 for the prisoner's dilemma, then analyze how the boys' decision making should proceed.

4. Employment negotiations often become contentious over the parties claiming individual value, such as salary. What are some issues to focus on during employment negotiations that will create joint value?

10

Project Management

10.1 Introduction

Previous chapters have focused primarily on technical aspects of analytics; this chapter and the one following it will concentrate on two critical "soft skills."[1] In this chapter, I will discuss the manner in which analytical work is organized and how project management can be used to make that organization more efficient. Project management is essentially analytics applied to the act of working; commonly used in a wide variety of fields, it improves gut-based project organization through the use of structured techniques.

Some types of analysts, such as business analysts, have project management well-integrated into their educations, but other types of analysts, such as scientists, generally undertake no formal study of the subject and instead rely on learning how to manage projects only from experience. In my opinion, that lack of formal training is a serious oversight. There are real benefits to using structured project management techniques and it's not as if they are hard to learn or to implement. It's also helpful for all the team members, not just the project lead, to understand the basics of project management. Right now, the business analysts who already are familiar with these techniques are nodding their heads.

What do I mean by project management? First, a project is a set of activities carried out to achieve a goal. Projects have start and end dates; work that goes on indefinitely isn't generally called a project. Notice how this definition encompasses many types of analytic work. Project management is a group of structured techniques that are used to organize the completion of a project and aid in the efficient application of resources. In project management, resources are anything used to complete a project, including personnel, money, time, and special equipment. Typically, there are restrictions on the resources that can be applied to a particular project – nobody has infinite personnel, money, or time. The project lead is the person ultimately responsible for the success of the project.

When used properly, project management has many benefits. It helps the project lead anticipate challenges so he or she can prepare resources ahead of time and isn't taken by surprise. Additionally, it provides a systematic

---

[1]To many analysts, the term "soft skills" has an unfortunate derogatory connotation.

method with which to forecast completion dates – furthermore, if a project is going to run over time, project management helps the lead determine where the application of more manpower or money can help. Project management is especially useful when a project has interdependent and overlapping steps or when a group of analysts must coordinate to get the work done. If you think that project management is best done "by eye" without an analytical approach, ask yourself how many of your technical analytical findings were only revealed through structured approaches.

In this chapter, I'll give a brief overview of project management and describe a few simple examples. By no means is this a comprehensive treatise on project management; in fact, the field is rich enough that there are textbooks written on individual subfields. If you'd like to learn more about the subject, start at the professional society for project managers, the Project Management Institute (www.pmi.org). You'll find additional references and also a path to professional certification.

As you read this chapter, think about how these techniques apply to your own work. Project management is really learned through application, rather than by just reading about it. I promise you that you'll reap real benefits if you implement these ideas.

10.2   How Analytic Projects Are Typically Managed

Let's discuss how analytic projects are typically managed – in short, they are not. Gut-based management is widespread. For example, new assignments are often given to the analyst who appears to have the most free time without thought to the assignment's interaction with the rest of the project. Project completion dates are plucked out of thin air, and when part of a project is late the lead panics and throws more money and people at it, hoping they will help.

In many cases, these decisions are made because the project lead simply doesn't recognize the benefits of using structured management and planning techniques and makes resource decisions using his or her gut. In other cases, the neglect of project management is a result of the short time frames that analysts often operate under. When a decision maker demands analysis to support a decision, it can be difficult to resist starting analysis immediately to instead spend time early in the project to plan. After all, most of an analyst's education is focused on analytical techniques, not managing people and resources – that management stuff is simple, right?

Wrong. In my experience, the success or failure of a project relates much more strongly to aspects of management than to the project's technical aspects. Furthermore, even in successful projects, a haphazard approach to project management can lead to an inefficient application of resources. For example, personnel may have to wait to use limited resources, such as specialized equipment, when the hold could have been avoided if parts of the project were completed in a different order. The completion of the project is then delayed, costing extra money and effort.

10.3   A Methodology for Project Management

Project management is a technique that can help minimize haphazard management and the inefficient use of resources.

There are two technical terms of art I will need to use here, so let's pause and quickly define them. Tasks are the activities that, when grouped together, make up the work of a project. The critical path is the chain of tasks that constitute the longest duration through the project. I'll discuss this idea in more detail later, but for now keep in mind that the critical path is used to determine when the project will be complete.

In this chapter, I'll describe a methodology for project management that consists of the following steps:

1. Define the project's goal.

2. Delineate roles.

3. Determine the tasks to achieve the goal.

4. Establish the dependencies between tasks.

5. Assign resources to tasks and establish a schedule.

6. Execute and reevaluate.

The first five steps of this methodology pertain to planning the project, rather than executing it. Many project leads document the planning phase with a written project plan, which contains all of the information gathered during these steps.

Even for the most straightforward projects, it is rare to proceed through this methodology linearly. Often, analysts must return to earlier steps and repeat them in light of new information. For example, when a decision maker and a project lead come to agreement on a project's goal, the lead may put together a project plan and find that, given the resources at his or her disposal, the estimated completion date is unsatisfactory to the decision maker. The project lead can then ask for more resources, revise the project's goal in consultation with the decision maker, or abandon the project entirely. The first two of these choices require redoing steps in the project management methodology and building a revised project plan.

Let's discuss each step in the methodology, in turn.

10.3.1   Define the Goal

The first step in the methodology, defining the project's goal, is important in setting the stage for the work of the project. The project's <u>goal</u> is the reason the project is being embarked upon; it is the product, support, service, or effect that will be delivered or accomplished when the work of the project is done.[2] Some examples of project goals are:

- This project will supply the CEO with an accurate and up-to-date cost benefit analysis to inform his decision regarding whether to purchase or lease a new office building.

- This project will provide a better understanding of the mechanism by which a specific gene influences whether an individual develops a disorder, documented in a research paper.

- This project will provide analytic support to my decision of which car to purchase.

- This project will move my household from its current apartment to a new apartment in an as yet undetermined location in town.

I will refer to the last of these goals as the "moving project" and use it as an example throughout this chapter.

At the start of a project, a succinct statement of the project's goal, similar to these examples, should be negotiated and agreed to by everyone participating. For an analytic project, it is especially important that the ultimate end user of the analysis agree to the goal. My primary concern when defining a project's goal is controlling the expectations of this end user, because if the expectations for the end product are unrealistic and don't match with what I could conceivably deliver, then the entire project may be a waste of resources. Some project management methodologies formalize these expectations and roles using a project charter, which is agreed to by all the participants; I find that charters are most useful when team members have contentious relationships, but otherwise they can be a little pedantic. In any event, I would recommend writing down an agreed-upon version of the goal, even if everyone's agreement is kept informal.

In addition to discussing what is part of the goal, all the participants should also agree as to what is not included in the project. The combination of what is in and what is out is known as the project's <u>scope</u>. For example, consider the moving project: is unpacking at the new apartment considered part of the project? What about painting the new apartment? These types of discussions should occur at this stage. However, in practice, a decision maker's expectations will often expand as the project progresses and meeting these new expectations will require more work. Project managers refer to this

---

[2]Many authors refer to the goal as the project's objective, but I have reserved that term for other uses in Chapter 9.

behavioral pattern as scope creep; as it disrupts plans and schedules, it is typically something they try to avoid.

Many project leads find it helpful to create objective metrics that can be used to measure the success of the project. These metrics allow for performance evaluation, after all the work has been done. Common metrics pertain to the cost of the project, the quality of the analysis delivered, and the timing of when the project is complete. The metrics don't have to be quantitative if there isn't a readily available scale; building a constructed ordinal scale, as described in Chapter 2, is a viable choice. Make sure that the metrics are understood by all those participating in the project, especially the ultimate end user of the analysis – the conversations around the metric will help the project lead understand the end user's expectations.

When defining the goal, it is helpful to discuss a few related aspects of the project's context, such as assumptions and constraints. Assumptions are conditions that are generally outside of the team's control and must be true for the project plan to be accurate and, in some cases, for the project to be successful. For the moving project, an assumption could be that all of the belongings will fit into a single truckload, meaning multiple trips back and forth between the apartments will not be required. A constraint is a restriction on a resource (be it time, money, personnel, etc.) that the project must obey; constraints were discussed in the context of decision alternatives in Section 9.2. The assumptions and constraints should be listed and documented in the project plan, and revisited and updated throughout the course of the project.

10.3.2 Delineate Roles

When working with larger teams (more than four people), it is useful to discuss and formalize roles amongst the team members at the start of the project. A project lead, who has accountability for achieving the project's goal, authority to make decisions, and responsibility for communicating with the end user, is a key role. The project lead often delegates some of his or her responsibilities to other team members. For example, a team member with knowledge of accounting could be delegated the authority for making day-to-day financial decisions, or a team member with project management skills could be delegated authority for day-to-day management decisions if the lead is busy completing other tasks.

As the project roles are assigned, it is imperative to establish how team members will communicate with each other. Discuss both the organizational aspects of communication (e.g., the paths by which team members pass information, what type of information should be shared with the whole team, and what sorts of meetings need to be held) and the more mundane aspects (e.g., who should be cc'd on emails, the rate with which updates need to be sent, and the frequency and location of meetings).

10.3.3 Determine the Tasks to Achieve the Goal

Tasks, also known as activities, are the building blocks from which projects are assembled. A single task can be worked by one or more people, depending on the nature of the task. The project lead tracks the completion of tasks to monitor the progress of the project. For this reason, tasks vary in size from project to project; if I were building a project plan for a five-year-long project involving hundreds of people, the tasks in that project plan would be larger units of work than if I were building a project plan for a month-long project involving two people.

At the end of this step in the project management methodology, the project lead should have a list of all the tasks that must be completed to successfully accomplish the project's goal coupled with an estimate of the amount of effort required to complete each task. When building a task list, the relationship between the tasks (such as which must come first) doesn't matter; those relationships are going to be captured in a later step in the methodology. It is possible to include optional tasks for things that are "nice to have" but aren't absolutely necessary for the project to be completed. These tasks could be pursued if there are spare resources available at some point in the project.

To demonstrate how this step works, let's use the example of moving to a new apartment. There are a number of tasks that are included within this project: packing belongings, searching for a new apartment, signing a new lease, renting a moving truck, etc. Some of these tasks take longer to accomplish than others, some can only be done by the person who is moving, and some can be finished more quickly if friends pitch in.

Writing the list of tasks that make up a project can be done by just thinking about the project, but using a structured technique increases the likelihood of writing down a complete list. One appropriate technique is called a <u>work breakdown structure</u> – a hierarchical treelike diagram that depicts how the overall work of the project breaks into tasks. The top node of the diagram describes the project as a whole, and each layer underneath breaks the work into smaller and smaller parts. The bottommost layer is generally the task level, though there are times when it is useful to go into further detail and list subtasks.

Figure 10.1 depicts a work breakdown structure for the moving project. I first broke the project down into three parts: moving out of the old apartment, transferring possessions to the new apartment, and finding and readying the new apartment. Each of these parts is broken down further into multiple tasks. Note that because of the way the goal is defined, I haven't included unpacking possessions at the new apartment or returning the moving truck as part of this project; essentially I've made the determination that these tasks are out of scope. These types of decisions will come to the forefront at this stage, if they haven't already.

Some of these tasks can be completed more quickly with added manpower, and some of them take the same amount of time regardless of how many people

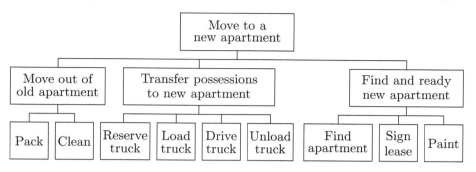

FIGURE 10.1
Work breakdown structure for the moving project. The project breaks down
into nine tasks at the bottom level of the hierarchy.

are working on them. For example, if it would take a single person 30 hours to
pack all their belongings, getting help from two friends could conceivably cut
that to 10 hours of time (even though it is still 30 total hours of labor). On the
other hand, no matter how many friends I ask to come sign a new lease with
me, it's still going to take me 2 hours. These two types of tasks are known
as effort-driven tasks and fixed-duration tasks. The task list should include a
note regarding the category each task falls under.

In the simplest case, effort-driven tasks require just a total number of man-
hours to complete, but there are some limits to this task model. For example,
consider the task of painting the new apartment, which I've estimated takes
20 man-hours to complete. Two people could conceivably finish the work in
10 hours, but if I had 40 people to complete this task there's no chance it
would only take half an hour. For some tasks, there are a maximum number
of people that can productively participate, and beyond that number more
people just are not helpful. Additional people also come with a requirement
for increased communication and planning; this resource cost will be discussed
in more detail in Section 10.4.

How can the project lead estimate the effort required to complete a task? In
many cases, the lead has experience working on similar projects and uses that
background to derive these estimates. Estimates can also be gathered from the
project team members who will be carrying out the work; being closer to the
task, they often have a good understanding of the labor involved. In any case,
both the lead and the team members should agree on the effort estimates,
otherwise conflict is the likely result.

Note that it is possible to include uncertainty in these effort estimates;
the Program Evaluation and Review Technique (PERT) developed by the
U.S. Navy is an example of a project management methodology that incorporates uncertainty straightforwardly (Miller 1963). PERT depends on three
estimates of the effort to complete a task: the optimistic estimate (W_O), the

TABLE 10.1
Task list for the moving project. Hours of effort are given for a single individual to complete the task. Tasks are categorized as effort driven (E) or fixed duration (F).

Task Name	Hours of Effort	Type
Pack belongings	30	E
Find a new apartment	20	F
Sign a new lease	2	F
Reserve a moving truck	2	F
Load moving truck	4	E
Drive truck to new apartment	1	F
Unload moving truck	4	E
Paint new apartment	20	E
Clean old apartment	6	E

pessimistic estimate (W_P), and the most likely estimate (W_L). The methodology then combines these estimates into a composite estimate (W_C) via a weighted mean:

$$W_C = \frac{W_O + 4W_L + W_P}{6} \tag{10.1}$$

The composite estimate is then used instead of the most likely estimate going forward. This is a simple way to incorporate uncertainty, but it does leave something to be desired, since Equation 10.1 simply approximates a distribution with a single number.[3] One alternative is to carry forward multiple estimates of the effort required to finish a task; this is especially appropriate when most of the uncertainty is concentrated in a single task. Another alternative is to use the event chain methodology, discussed briefly in Section 10.6.

Why does breaking a project down into tasks and estimating the effort required to complete them individually result in a more accurate estimate than examining the entire project at once? Decomposing a question into smaller connected estimates is, in general, more accurate than trying to incorporate all the expert's experience into a single estimate (Hora 2007).

In Table 10.1, I've taken the tasks from the work breakdown structure for the moving project and created a task list. Along with the name of the task, I've written an estimate of the number of hours of labor necessary to complete each task, as well as whether it is effort-driven or fixed-duration.

---

[3]To be fair, the original version of the PERT methodology did also include an estimate of the effort's variance: $s = (W_O - W_P)^2/36$.

Tasks with Drastically Uncertain Durations

In analytic projects, many tasks have uncertain durations. I'm not referring to garden variety uncertainty along the lines of "It will take me somewhere between 10 and 15 minutes to wash the dishes." Instead, I'm talking about deep uncertainty that can span several orders of magnitude. Tasks with this quality can often be thought of as puzzles – there is a solution possible, but there is no good way to estimate how long it will take someone to find it. Some piece of creative inspiration is necessary before the task can be completed, and that inspiration could arrive in 10 minutes or it could take a whole year. Even worse, that inspiration may never arrive. Hunting for bugs in a piece of code is a common example of this type of task.

Project management cannot change the nature of these tasks, but it can help manage their completion and mitigate the task's impact on the rest of the project schedule. It is of the utmost importance to communicate to the other people involved with the project the level of uncertainty associated with the length of the task. This helps to level expectations and budget resources. If you are the project lead, map out best- and worst-case scenarios for resource allocation before things come to a crisis point.

I suspect that the common appearance of highly uncertain tasks in analytical work is another reason many analysts do not use project management techniques. It is not pleasant to be held accountable for a task taking too long if the estimate for how much effort it would take to complete was so deeply uncertain in the first place. In my view, forgoing project management techniques for these tasks is still a poor choice, because when you miss a deadline or run over budget the other team members are caught by surprise and must improvise a solution. In other words, avoiding project management techniques will not change the nature of the task.

10.3.4   Establish the Dependencies between Tasks

There is some flexibility in the order in which the tasks that make up a project can be completed, but they generally cannot be completed in any arbitrary order. For example, some tasks need to be completed before others can be begun. The relationship between two tasks is known as their dependency. Studying these dependencies is essential for the project lead to develop a viable order with which to complete the tasks. Two tasks don't necessarily need to have a dependency; they can be unrelated.

If two tasks are connected, their dependency can be placed into one of the following four categories (though, of course, there are some grey areas):

- **Finish to start**
  Tasks that are finish-to-start dependent must be performed sequentially. In other words, the first task must be finished before the second can be begun,

though unrelated tasks can occur in between. For example, I can't load my moving truck before I finish reserving the truck; the rental company would certainly object if I tried to do this. This task relationship also occurs if two tasks require the same resource and that resource can only be used for one task at a time. For example, my friend has to get off the exercise bike before I can start using it.

- **Start to start**
  Tasks that have a start-to-start dependency occur when the first task must begin for the second to start. Simultaneous starts are also allowed. An example of tasks related in this way are the editing and writing of a document. I often edit the beginning parts of a document while I'm writing the later sections. Thus, in order to start editing I must start writing first – otherwise there wouldn't be any text to edit.

- **Finish to finish**
  When one task must be completed for the second to be finished, the tasks are finish-to-finish dependent. This dependency often occurs when one task is a part of a larger task. For example, in order to finish cleaning my apartment, I must finish washing the dishes. This relationship can also occur when two tasks are proceeding concurrently in an assembly-line manner. For example, two people can pack boxes and load a moving truck concurrently, but to finish loading the truck the packing must also be finished.

- **Start to finish**
  Sometimes, to finish a task, some other dependent task must be started. This is known as a start-to-finish dependency. As an example of this type of task, imagine you are a caregiver for a sick patient. In order for you to finish your day's work, the next caregiver must arrive and start their shift. The overlap in the two tasks can be minimal or the first task can be finished while the second simultaneously begins, but there cannot be a time when neither caregiver is on duty.

There is no limit on the number of other tasks that a task can have a dependency with. For example, in the moving project the packing task has finish-to-finish relationships with both loading the moving truck and cleaning the old apartment.

How does a project lead actually determine the task dependencies? There isn't a shortcut for this process; the lead must proceed through each task and think about which other tasks it is related to.

Once you have determined the dependencies between your tasks, you can represent those relationships graphically using a network diagram. Arrows connect dependent tasks, with the arrowheads pointing to or emerging from the left side of a task for the "start" part of a dependency and the right side for the "finish" part of a dependency. Working from a network diagram, it is easy to determine viable orders in which to complete tasks. Just follow the arrows and make sure that all of the dependencies have been satisfied before

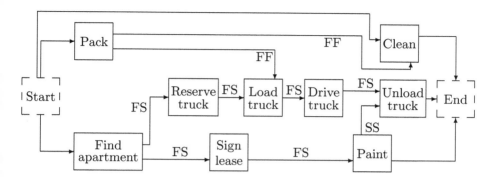

FIGURE 10.2
The network diagram for the moving project. Tasks are in solid boxes and the start and finish of the project are in dashed boxes. The dependencies between the tasks are marked with arrows, and each dependency is labelled with its type, including finish-to-start (FS), start-to-start (SS), and finish-to-finish (FF).

proceeding. Note that the drawing of a network diagram and the relationships between the tasks should be independent of resources. In other words, it doesn't matter how many people you have packing your belongings; you still need to finish packing before you can finish loading the truck.

Consider Figure 10.2, the network diagram for the moving project. Tasks are in solid boxes, with the start and end of the project in dashed boxes. Arrows connect dependent tasks, with labels on each arrow depicting the nature of the dependency. I've distributed the tasks vertically on the diagram corresponding to which of the three parts of the project they belong to (see Figure 10.1). However, in a network diagram the spatial placement of the task boxes doesn't matter; only the arrows connecting the tasks contain information.

It's worth briefly discussing the dependencies in the moving project network diagram, because some of the choices I've made regarding the type of dependency are debatable. For example, I've shown painting the new apartment and unloading the truck as start-to-start dependent; I did this because I like to have one room where painting is complete to unload the boxes. Some people don't have that requirement and would draw their network diagram differently. Look through the relationships I've drawn and see if you agree with the dependencies I've chosen.

There are three tasks available to be worked on from the start of the project: packing, cleaning, and finding an apartment. Cleaning the old apartment cannot be completed until packing has finished. Once the find apartment task is complete, two more tasks become available to be worked: reserve the truck and sign the lease. The rest of the tasks can proceed according to the dependencies in the network diagram.

Some project managers choose to build <u>milestones</u> into their project plans at this point. Milestones are significant project achievements that demonstrate how far the project has progressed to the team and to others and are often causes for celebration. Typical milestones signify the transition from one phase of the project to another (e.g., from designing to building a new product). For the moving project, a useful milestone could be the completed signing of a new lease, which demonstrates the commitment to vacate the old apartment and marks a point of no return.

10.3.5 Assign Resources to Tasks and Establish a Schedule

The final planning step in the project management methodology before execution begins is the assignment of resources to tasks, and the establishment of a schedule based on those assignments. This step is where all of the planning work comes together.

Project management has a specialized graph commonly used for scheduling known as a <u>Gantt chart,</u> named after Henry Gantt, an early innovator in management science. In a Gantt chart, tasks appear in a vertical list, with time proceeding horizontally from left to right. Horizontal lines across from each task name mark the time when the corresponding tasks are being worked. The arrangement of these lines in the Gantt chart must obey the dependencies in the network diagram and the duration estimates in the task list. The project's completion date can be estimated from when the last task is scheduled to be finished.

Many variations on the simple Gantt chart exist. For a project in progress, a vertical line is sometime used to mark how far work has proceeded. Some project managers like to draw the dependencies from the network diagram directly onto the Gantt chart. Milestones are also commonly marked. For projects involving multiple workers, it can be helpful to label each line with the number of people working the task.

Depending on how the available resources are assigned to tasks, many viable Gantt charts are possible for each project. For example, for the moving project the choice of which task to execute first impacts the Gantt chart.

The term critical path is used in project management to refer to the longest chain of tasks in the project. The critical path determines how long it takes to complete the project. The tasks that make up the critical path have no <u>slack;</u> that is, they proceed in sequence without any gaps in times between the end of one task and the start of the next one.

Figure 10.3 shows a possible Gantt chart for a single individual completing the moving project. The individual proceeds through each task in sequence, starting at hour 0 of the project by working on the find apartment task and reaching project completion at hour 89 when cleaning the old apartment is finished. This is one viable order in which the tasks can be completed, but

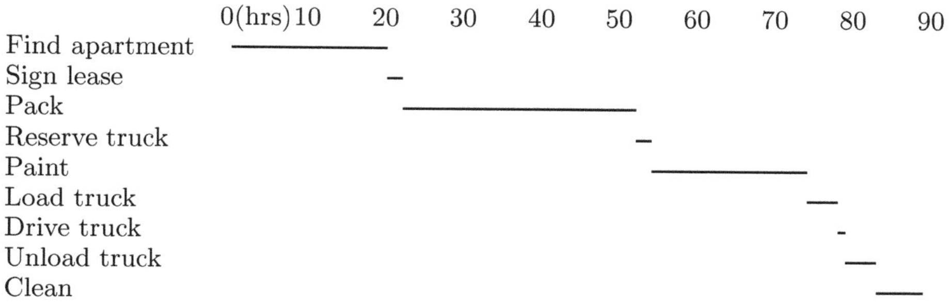

	0(hrs)	10	20	30	40	50	60	70	80	90
Find apartment	───────									
Sign lease			─							
Pack				─────────					
Reserve truck							─			
Paint								───────	
Load truck									──	
Drive truck									-	
Unload truck									──	
Clean										──

FIGURE 10.3
Gantt chart for a single person doing all the work of moving to a new apartment.

	0(hrs)	10	20	30	40	50
Find apartment	────────────				
Sign lease			─			
Pack	──────────────				
Reserve truck			─			
Paint			────────────	
Load truck				──		
Drive truck				-		
Unload truck				──		
Clean				───	

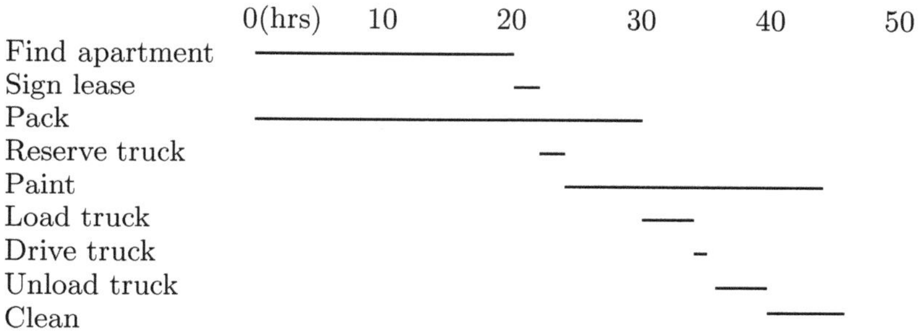

FIGURE 10.4
Gantt chart for two people working on moving to a new apartment, but assigned to tasks independently.

many others are possible as long as the order obeys the task dependencies. However, with a single individual working the project, it will always take 89 hours of work to reach completion, because that's how much work there is to do (ignoring uncertainty in the task duration estimates).

When a second individual is added to the team, things get more interesting. In Figure 10.4, I depict a viable Gantt chart for two workers. In this example, the project is planned such that the individuals do not work cooperatively on any single task, but instead simply divide the tasks up. Here, the first individual completes the following tasks in series: find an apartment, sign a lease, reserve truck, and paint. The second individual packs, loads the truck, drives the truck, unloads the truck, and cleans. The project is completed in 45 hours, when the second individual finishes cleaning. With this allocation of

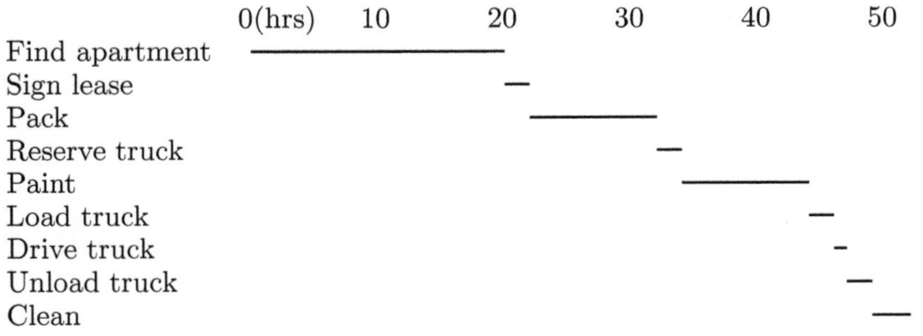

	0(hrs)	10	20	30	40	50
Find apartment	————————					
Sign lease			—			
Pack			————			
Reserve truck				—		
Paint				————		
Load truck					—	
Drive truck					-	
Unload truck					—	
Clean						—

FIGURE 10.5
Gantt chart for two people working together on every task for moving to a new apartment.

resources, the series of tasks completed by the second individual is the critical path.

Figure 10.5 depicts a Gantt chart for two individuals collaborating on every task. This only reduces the time to completion for the effort-driven tasks; the fixed-duration tasks, unsurprisingly, take the same amount of time. This means that for fixed-duration tasks, there are more man-hours being expended than the minimum necessary to complete the task. That isn't necessarily a bad thing, but it's apparent that it impacts how long the project takes to finish; the Gantt chart shows that it takes more than 50 hours from beginning to end.

When using a Gantt chart to communicate a project schedule, be aware that they have several limitations. These limitations include:

- Gantt charts don't convey the percentage completion of each task for a project in progress. For example, a task could be only 33% complete, but have used up 50% of its allotted time.

- A delay in one task causes ripple effects in other tasks that are not immediately apparent on a Gantt chart. This often happens because personnel have to stay on the task longer, preventing them from moving to their next task, but the tasks a team member is scheduled to work next aren't marked on the chart.

- Gantt charts don't include uncertainty in the estimates of how long it will take to complete a task.

- Projects that have a large number of tasks with many resources assigned are difficult to follow on a Gantt chart, especially for viewers not intimately familiar with all the tasks.

10.3.6   Execute and Reevaluate

Once the planning phases of the project are finalized and agreed upon by everyone involved, it is time to start executing the project itself. Progress in completing the tasks can be marked on the Gantt chart via a vertical line labelled "today."

While executing the project, the project lead will almost always find that things haven't gone according to plan. Either a task takes longer than anticipated, a key piece of equipment breaks down, an assumption turns out not be true, or an essential team member gets sick. This stuff happens – using project management doesn't prevent bad things from happening, but it makes these bad things easier to work around.

When the execution of the project starts to differ significantly from the project plan, it is time to revise the project schedule and the accompanying Gantt chart. For tasks that are in progress, adjust the length of the task to reflect the updated knowledge on how long they will take to complete. Do the same for tasks that have already been finished, so that the Gantt chart reflects the project as it has come to pass. Adjust the future assignment of resources so that they continue to be allocated in an efficient manner.

After a project has been completed, it is important for the participants' professional development to complete a lessons learned report. A lessons learned report documents both the positive and the negative takeaways gathered during the life of the project. Lessons learned reports can be created for individuals involved in the project, especially the project lead, and for the team as a whole. Reflecting on a project and recording these lessons learned is essential for individual growth as an analyst and for the team's growth as well. As an example of a lesson learned for the moving project, consider the following: "Moving truck rental companies will often not have a truck available at the correct time even if a reservation has been made. Team members should have a contingency plan for use if the truck is not available at the scheduled time."

10.4   Hastening Project Completion

One of the most common complaints the project lead will hear from the end user is that the completion date is not soon enough. How can the lead speed up the completion of the project?

If the lead wants to shorten the overall length of the project by adding resources, they should be added to one or more of the tasks on the critical path, because adding resources elsewhere will not impact the length of the overall project. It's possible this addition of resources can change which tasks constitute the critical path; when that happens, to reduce the time to completion further, resources should be added to the new critical path. The lead also has the option to rearrange the order in which tasks are completed and how the existing resources are applied to those tasks, but I'm assuming the resources at hand have already been optimally assigned.

Adding resources to an effort-driven task does not always result in that task being completed more quickly. This is especially true for new personnel added to an existing team, because when multiple people are staffed on as task or a project, the effort necessary for communicating and coordinating their efforts acts as a tax on their time. Consider Brooks' Law (Brooks 1995):

"Adding manpower to a late software project makes it later."

How can this be possible? Shouldn't more manpower cause the work to get done faster? Not necessarily; new workers must be trained, someone must manage them, information needs to flow to and from them, and the tasks may need to be redivided so they have a cohesive piece of the project to work on. For these reasons, don't assume that adding manpower to a task will help it finish more quickly. Talk to the people working on this task and ask them to estimate how much effort will be required to bring new workers up to speed, the additional effort that will be necessary to coordinate with them, and whether there are nicely packaged tasks or subtasks that can be handed off to them.

Perversely, when totaled over the length of the project, slowing down a project's completion also costs more resources! Salaries, benefits, and rent must be paid for longer periods of time, people with essential expertise leave and must be replaced and people take some time to get back up to speed on a project if they've been multitasking and working on other things. Essentially, changing your plans, whether lengthening or shortening the timeline, will cost extra resources. This state of affairs led Norman Augustine to posit his Law of Economic Unipolarity: "The only thing more costly than stretching the schedule of an established project is accelerating it, which is itself the most costly action known to man (Augustine 1997)."

10.5 Project Management of Analytic Projects

It's easy to see how to apply project management to physical projects, like moving from one apartment to another or the construction of a building. The application of these techniques to analytic projects can be more difficult to envision, so let's illustrate it with an example.

Imagine you are leading a small team of four analysts charged with providing quick turnaround analysis to a decision maker. Let's not worry about the subject matter of the decision and just focus on the process of how the project plan is constructed and the project is completed. The project begins on Monday morning and must be complete by close-of-business on Friday, at which point the decision maker will use the analysis to make her choice. For planning purposes let's say that each analyst works five 8-hour days.

The project begins with all members of the project team meeting with the decision maker to hear her articulate the kind of information needed for the decision. The scope of the project and the context for the decision are important parts of this discussion. Afterwards, the lead convenes a team planning meeting to discuss team roles, break the work into tasks, develop a schedule, and cover all the other topics that would be included in a project plan. With a turnaround this quick, a formal project plan may not actually be developed, but all of the topics that would be included in that plan should be discussed and agreed upon.

At the planning meeting, the team breaks the project into a few major tasks, with some time for communication and coordination also allocated. The first phase of the project consists of reviewing prior work and gathering data specific to the decision at hand. These two tasks can proceed concurrently, but they both have a finish-to-start dependency with the task that begins the model-building phase – a methodology meeting. To inform this decision, two models will need to developed; the models will be integrated into a joint analysis to inform the decision. The development of the models can proceed concurrently, but the finished models must be integrated into a combined model. Figure 10.6 is a viable Gantt chart for this project, which I will now discuss in more detail.

The first two days of the Gantt chart are dedicated to research and data gathering. Day 1 begins with the two meetings mentioned previously. The lead splits the analysts into two groups of two, with one group reviewing prior work on similar decisions and the other group gathering data specific to this decision. The team estimates that these tasks will take the groups the remainder of the first day and the majority of the second day to finish. This work is scheduled to be interrupted on the morning of the second day by a half-hour coordination meeting wherein each team presents the results of their research to that point. The team will discuss any potential problem areas with these first tasks, and afterwards return to their work with suggestions from other team members as to how to address the problems. The second day closes with a methodology meeting during which the two models and the plan for incorporating the data and prior work are discussed.

The third day begins with a "touch-base" meeting with the decision maker, where the key findings of the research are presented and the proposed methodological approach is discussed. The decision maker then has a better understanding of the analysis she will be receiving and can give approval for the next phase of the project to proceed; the team can think of this approval as a milestone. After a coordination meeting, the remainder of the day is devoted entirely to the independent development of the two models. The team will again be split into two groups, each responsible for the building and testing of one of the models. In this Gantt chart, the team has estimated that the second

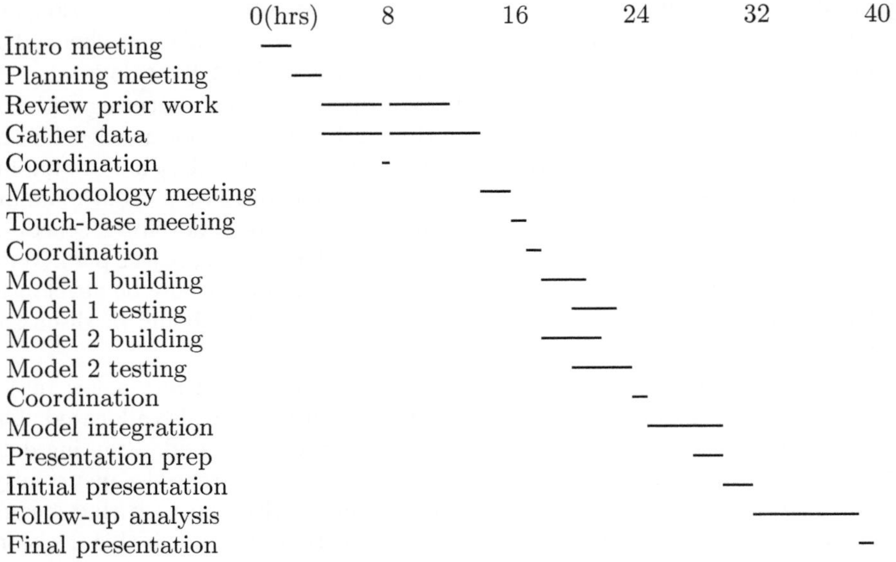

	0(hrs)	8	16	24	32	40
Intro meeting	—					
Planning meeting	—					
Review prior work		— —				
Gather data		— —				
Coordination	-					
Methodology meeting			—			
Touch-base meeting			—			
Coordination			—			
Model 1 building				—		
Model 1 testing				—		
Model 2 building				—		
Model 2 testing				—		
Coordination				—		
Model integration					—	
Presentation prep					—	
Initial presentation					—	
Follow-up analysis					—	
Final presentation						—

FIGURE 10.6
Gantt chart for the analytic project example.

model requires somewhat more effort than the first one. For both models, note that the testing and building tasks overlap, as the analysts anticipate testing the smaller pieces of the models as they are built.

The fourth day opens with a coordination meeting, where each team presents their model. The team then works together to integrate the two models into a single piece of analysis, with one analyst breaking off partway through this task to develop the presentation material. The model integration and the presentation preparation tasks have a finish-to-finish dependency. The last task of the day is the presentation of the initial analytic findings to the decision maker.

The team anticipates that the decision maker will have follow-up questions during this presentation, so the lead has budgeted the fifth day for additional analysis and a final presentation. When the project plan is being created, the team doesn't yet know what follow-up analysis they will need to conduct, so there is some risk in leaving only a day to perform that analysis. This illustrates the importance of the touch-base meeting on the third day, to ensure that the team is on the right track.

A more detailed project plan could include tasks of a more granular detail, but with only four analysts staffed on a weeklong project, this level of detail is usually sufficient. Of course, there is no guarantee that the actual project will play out this way. There is uncertainty in all of the task duration estimates, and if one of the tasks runs long it will require the lead to reallocate resources. However, with a project plan and Gantt chart in place, the lead can easily see the impact of a task running over time and where resources can be reallocated from.

10.6 Other Project Management Methodologies

The project management methodology I described in this chapter is known as the "traditional approach," and I've also mixed in some of the critical path method. A variety of other methodologies also exist – often they've been optimized for specific environments or to emphasize particular advantages of project management. Here are just a few additional varieties:

- Critical chain project management combines aggressive scheduling of the individual tasks and the accumulation of a resource buffer at the end of the project to better ensure on-time project completion (Goldratt 1997).

- Event chain methodology is a project management technique used to model variances in the time to complete tasks – the modeling is usually accomplished through probabilistic simulation built into a software program (Virine and Trumper 2008).

- Agile project management is particularly popular amongst software developers. It emphasizes flexibility over the rigidness of the traditional approach and frequent product releases with iterative improvements (Cohen et al. 2004).

10.7 Exercises

1. You are the lead for a project in which the goal is to provide the decision maker with a cost-benefit analysis regarding whether to purchase or lease a new office building. You have been asked to put together a team of four analysts (who will report to you) to produce the analysis. After the analysis is complete, it must then be presented to the decision maker. What skills would you look for to fill each position? Building a work breakdown structure may help you answer this question. How would you ask your team to communicate?

2. Imagine that you are considering looking for and accepting a new job. What tasks make up this project? Assemble them into a work breakdown structure. What are the dependencies between the tasks?

3. Consider the moving to a new apartment project management example. Create a Gantt chart assuming that yourself and two of your friends are working on this project. What is the quickest time to project completion? What is the critical path?

11

Communicating Analytic Findings

11.1 Introduction

The last major topic in this book and the second of the "soft skills" I will cover is the communication of analytic findings. History is littered with analysts who found signals in data, but were unable to convey the importance of these signals to people who could act on them. If you want your work to have impact (and you aren't running the show by yourself), communicating analysis is an essential skill to master.

Davenport et al. (2010) argue that communicative abilities are part of a larger set of "relationship and consulting skills" that analysts use to work with their colleagues and customers. In their words, "effective analysts need to be proficient not only with data, but also with people." Communication is an essential part of every step in the analytic process from design to implementation.

Communicating analytic findings well requires expertise in several different, but related, capabilities:

- Explaining technical concepts using both the written and spoken word,

- Building graphics that clearly present findings and ideas (a subject known as data visualization), and

- Tailoring a message to particular audiences.

Analysts aren't born with these capabilities, Fortunately, they can be learned and improved through practice and training.

Of all the topics we've discussed in this book, analytic communication is changing the most rapidly. Several recent technological developments, including the proliferation of analytic software packages and the rapid growth in computational processing and data storage, have made it possible for all analysts to process extremely large data sets and build custom visualizations. Complex visualizations are beginning to appear in mainstream media, such as newspapers and magazines as well as on popular websites. Analytic ideas are becoming a larger part of our professional, academic, and political discussion. The technological developments have led to a flourishing of new ideas, and even the evolution of a new method of data visualization: the interactive graphic.

This chapter is split into four parts. The first part describes basic principles of analytic communication. The second part discusses the media over which analytic communication takes place, providing specific advice for a few of the alternatives. The third part outlines a simple characterization of the variety of audiences for analytic communication, and how to tailor your message for each. The chapter concludes with some practical advice for visualizing data.

11.2 Fundamental Principles of Analytic Communication

Analytic ideas and findings are often surprising, subtle, and technically complex. These qualities can make them challenging to communicate, regardless of the audience. On the other hand, analysts have a great deal of freedom over the manner in which they communicate ideas and findings – some overarching, general principles can help analysts make decisions in this regard.

These sorts of principles are useful because, currently, communication advice for analysts is fragmented, primarily by medium. In other words, we use data visualization books to help us build plots, slide construction guides to help us build presentations, and web manuals to help us build websites. The advice specific to these media is very useful, but establishing overarching principles helps analysts make decisions regarding how to organize communication materials by keeping a small set of objectives in mind.

There are four principles that apply to all analytic communication, regardless of audience or medium. These principles are: clarity, transparency, integrity, and humility.[1] Whenever you are faced with a design decision for a communication product, return to these principles and they will guide you to a good solution. However, I don't mean these principles to imply that there is one right answer to communication decisions. Even with the constraints imposed by the principles there is still plenty of room for individual style, unique voices, and elegant solutions.

An alternative frame for these principles is to think of them as fundamental objectives (see Section 9.3) for the analytic communication process. Some alternatives will solely impact one of the objectives; for example, sometimes an analyst can improve the clarity of a plot by changing the color that the lines are drawn with. On the other hand, sometimes alternatives will involve tradeoffs between the objectives; those decisions are generally more difficult, and which alternative is preferred can depend on the audience or the medium.

Let's discuss each of the principles in more depth.

---

[1]A similar list of principles for excellence in analytic communication appears in Markel (2012).

11.2.1   Clarity

<u>Clarity</u> is the expression of an analytic concept in a manner that is simple, direct, efficient, and effective. Extraneous lines, words, colors, and other markings are minimized so the key idea can be placed front and center. At the same time, the concept must be expressed in a manner that is understandable and not oversimplified; minimization should not be taken so far that the finding disappears or key data features are lost.

Consider how experts on analytic communication in various media make recommendations that maximize clarity:

- In data visualization, clarity is exemplified by Tufte's first two principles in the theory of data graphics (Tufte 2001) – "above all else show the data" and "maximize the data-ink ratio." In other words, when making a data visualization, don't add tick marks, gridlines, or decorative elements unless they actually convey information. At the same time, don't eliminate any markings that impart information about the data.

- Iliinsky and Steele (2011), in their book on data visualization, are expressing the desire for clarity when they recommend "function first, suave second."[2]

- In his guide to slide presentations, Reynolds describes the Zen concept of simplicity as one of his guiding principles (Reynolds 2008). Reynolds' simplicity is similar to what I've called clarity, as evidenced by his advice that "simplicity can be obtained through the careful reduction of the nonessential," as long as it also "gets to the essence of an issue."

- In their classic book on style in writing, Strunk and White (1959) stress the importance of clarity by emphasizing the repercussions when an author fails to achieve it: "Muddiness is not merely a destroyer of prose, it is also a destroyer of life, of hope: death on the highway caused by a badly worded road sign, heartbreak among lovers caused by a misplaced phrase in a well-intentioned letter, anguish of a traveler expecting to be met at a railroad station and not being met because of a slipshod telegram. Think of the tragedies that are rooted in ambiguity, and be clear!"

- One of the rules of journalism is not to "bury the lede" – in other words, the writer should put the most important finding at the front of the story, stated directly. If the important finding is placed deeper in the story, the audience is more likely to miss it.

In summary, when communicating analytic ideas and findings, clarity means that you should maximize efficiency, whether measured through words, lines, or colors, while still conveying your thoughts forcefully, definitively, and, most importantly, understandably.[3]

---

[2]Personally, I would put suave even lower on the list.

[3]Scientists can think about clarity as maximizing the signal-to-noise ratio of communication.

11.2.2   Transparency

Transparent analytic communication explains to the audience the method by which the findings were derived, accessibly and to an appropriate depth. In addition to presenting the methodology, part of transparency is ensuring that the audience understands the assumptions that underlie the analysis. This is important because if the assumptions are violated (whether through further study, natural change, or some other means), or even if the audience just doesn't accept the assumptions, there will be implications for the findings. Transparency is most appropriately applied to the entirety of an analytic communication package, as opposed to a single plot inside of a technical document, for example.

Some examples of transparency in action include:

- A journal article that describes the methodology behind an interesting result. In many fields, the author of an article is required to publish his or her methods to the level of detail required for another researcher to replicate every step.

- A financial analyst making a presentation who discloses his data sources and the techniques by which he processed them.

- An analyst who publishes on the web the raw documents and code used in his new text mining method.

- A scientist speaking to the general public who, given the limited time allotted for her presentation, refers the audience to a freely available white paper for those who would like more technical detail.

Why is it important for analytic communication to be transparent? Shouldn't an analyst only care that the findings are communicated correctly?

First of all, one of the benefits of doing analysis is that there is a logical line of reasoning that leads to a finding; analysts don't need to rely on the assertion of findings without support. This gives analytics a competitive advantage versus other types of arguments, such as "gut-based" reasoning or subject matter expertise, and that advantage should only be squandered for very good reason. In other words, transparency builds the audience's confidence in the findings.

Secondly, part of our responsibility as analysts is to expose our line of reasoning to questions and comments. Sometimes this feedback reveals errors, oversights, or forgotten assumptions. These corrections should be welcomed, because in the long run they result in better analysis. In most cases, however, an analyst will have a ready answer to a question or a comment because he or she has spent more time thinking about the data and findings than the audience. Answering questions directly also increases the audience's confidence in the findings.

Finally, transparency helps to spur on other analysis. This occurs because revealing the methodology behind a finding can give a member of the audience an idea to solve a problem they've encountered or, even if transparency doesn't spark an immediate idea, it can add another tool to the analytic toolbox of the audience members. The audience can also more easily recognize other opportunities to apply analytics, sometimes bringing business and collaborative opportunities to the analyst.

The transparency communication objective demonstrates a benefit of keeping the line of reasoning as simple as possible – simple methodologies are easier to explain, and if you can't explain the methodology in a way that the audience will understand, few people will believe the findings.

11.2.3 Integrity

Analytic communication with integrity conveys defensible and sound conclusions driven by the data, while properly representing the limitations and uncertainty in the findings.

As analysts, our first responsibility is to ensure that we are communicating the data and findings accurately and honestly. However, it can be tempting to exaggerate the implications of the data, because, in all likelihood, no one will look at your data as thoroughly as you will. Analysts engage in exaggeration for many different reasons, whether to further their career, to please the audience, or simply to make a stronger argument than the data support. It is important to understand that this temptation is common; as analysts, we spend inordinate amounts of time and energy focused on our work and to be tempted by a larger reward is entirely natural.

However, it is counterproductive to engage in this kind of overreach, whatever the reasoning behind it. As discussed in Section 4.1, in the long run this kind of behavior will have a negative effect on your career, particularly in the opinion of other analysts. Additionally, as analysts, we study real phenomena, and our techniques are designed to reveal real insight regarding these phenomena. Even if your colleagues can't tell that you've gone too far, eventually the phenomena you are studying will show the truth.

In addition to communicating the data and findings accurately, the presentation of limitations and uncertainty in analytic communication is integral to integrity. This information allows the audience to use the findings responsibly – if limitations and uncertainty are not presented or are minimized, the audience is likely to apply the findings in regimes beyond which they are valid. It also facilitates comparisons between different analysts' findings.

Integrity is connected to the concept of "epistemological modesty," a complicated-sounding phrase that describes a simple idea. Roughly, analysts that demonstrate epistemological modesty do not overstate the findings and certainty of their work because they recognize that the real world is quite complex and often difficult to understand and model. As discussed in Section 4.6, analysis can break down in very surprising ways even if you've carefully accounted for the known sources of uncertainty. Keep this in mind when communicating findings.

A good example of the concept of integrity in action can be found in data visualization. When making plots, it is easy to exaggerate the trend you are trying to demonstrate by adjusting the axes in improper ways or by showing only selected subsets of the data.[4] Tufte (2001) expresses integrity by arguing that "graphical excellence begins with telling the truth about the data." In other words, the analyst should present the data in such a way that the audience leaves with an accurate and complete impression.

11.2.4   Humility

By humility in analytic communication, I mean that we should strive to remove the analyst from the message. In writing, Strunk and White (1959) recommend that authors "Place yourself in the background. Write in a way that draws the reader's attention to the sense and substance of the writing, rather than to the mood and temper of the author." In analytic communication, too often the audience takes away the idea that the analyst is some kind of super-genius, that analytical work is inaccessible, or that they could never carry out their own analyses. These perceptions are detrimental to the future of our profession; analytics is a young field and in order to grow we need to attract people and business by making ourselves and our work as accessible as possible. Furthermore, the data and the conclusions drawn from it should speak for themselves – if you find yourself needing to rely on your authority as an analyst, that's a sign that you may be overreaching.

We can communicate with humility by not encouraging a "cult of personality" around the analyst. For example, you can talk about mistakes that you made in the initial pass through the analysis or ways you feel the findings are difficult for you to understand. Discussing these sorts of things won't hurt the audience's opinion of you; in fact, it will actually improve it, because they will find you more relatable. Furthermore, they'll also think you are smart, in the way that we think great teachers are smart – good analytic communication requires a great deal of intelligence!

11.3   Medium

While the fundamental principles provide the overarching objectives for analytic communication, the medium through which communication occurs also influences how the findings are presented. The strengths and weaknesses of the

---

[4]This behavior is common in situations where the analysis was carried out to support a predetermined position – it's often seen in politics.

chosen medium, be it printed matter, oral presentation, an elevator speech, or web-based, need to be taken into account when designing the message. The analyst should anticipate the impacts of those strengths and weaknesses and tune the message accordingly.

11.3.1   Printed Matter

Printed matter has been used to communicate analytic findings for hundreds of years. It includes anything on the printed page, such as journal articles, technical papers, textbooks, magazine features, and other similar products. Printed matter is typically read by the audience while holding it at arms length or closer. It usually contains a mix of prose (in full sentences) and graphic components.

Communication using printed matter can take advantage of its many strengths. First, the resolution of the printed page ranges from good (newsprint) to excellent (magazine or computer printout). This resolution allows graphic elements to be reproduced with precision (though printing color in graphics can be expensive). Thoughtfully designed graphical depictions of data sets come across very well in printed matter, and when combined with a prose description of the graphic, they constitute a very effective means of analytic communication.

To be fully communicated and understood, analytic concepts often require equations to describe the technical details. Fortunately, printed matter is one of the few effective ways to communicate equations. The symbolic form of the equation can be interspersed with prose explanations of the equations' meaning.

Printed matter can also be written on, highlighted, and underlined by the audience. Prof. Eugene Chiang, one of my teachers, encouraged his students to "take notes in the margins of technical papers – that's what the margins are there for." This is good advice; annotating printed matter with your insights saves you time when rereading and reviewing old material.

Of course, printed matter has weaknesses as well. First, and most importantly, it is non-interactive. When the audience has questions about the material, there is no mechanism to interact with the author without turning to a different medium. Second, printed matter takes a long time to produce, especially when it contains analytic work. This is connected to its non-interactivity; writing comprehensive text takes a lot of effort and the text has to be complete enough that it answers the audience's questions. Finally, printed matter can be expensive to produce and distribute, especially for books and magazines. Depending on the type of printed matter, complex data visualizations with many colors can be expensive to print (and correspondingly expensive for the audience to purchase).

An additional characteristic of printed matter that has both positive and negative aspects is that there have typically been barriers to entry for authors. For example, journal publications and textbooks usually have to pass a peer review before being published to prevent incorrect material from appearing in print. These barriers also serve to protect publishing companies' investments – it is expensive to produce and distribute printed matter, and companies don't want to invest money in improperly written material and then fail to recover their investment. However, this barrier can also prevent materials with small niche audiences from being produced, because it is difficult to do so profitably. Thankfully, this effect has been minimized over the past few years, as many of these niche subjects can now be published on the web.

11.3.2   Oral Presentations

The most basic description of an oral presentation is a person speaking in front of an audience; for analytic communication, the speaker is often accompanied by projected or printed prepared slides, or alternatively a large writing surface (like a whiteboard or chalkboard) is available. Oral presentations come in many variants, including briefings, lectures, speeches, press conferences, and seminars. I'll discuss a specialized format, the elevator speech, in more detail in the next section.

The potential for interaction with the audience is one characteristic that differentiates an oral presentation from other media. Design your oral presentations to take advantage of this interactivity – be receptive to questions, allow audience members to express alternative points of view, and discuss those points of view when they are raised. Make sure to budget time in your presentation for this interaction, otherwise you'll feel rushed to get through the material and pressured to cut off discussion. If you budget too much time for discussion, it's not the end of the world, as no one ever complained about a presentation that ended early.

Whenever I give a presentation in front of a large group, I make sure to mention as part of my opening remarks that I want to be interrupted if any member of the audience doesn't understand any part of the material. It can be intimidating to ask a question in front of a lot of people, and I want to encourage everyone to speak their mind. In a small group I usually don't explicitly ask for questions because my tone is informal enough that no one feels any compulsions about interrupting.

One reason presenters should encourage questions is that when an audience member is confused about the material, it can distract them for the remainder of the presentation as they try to puzzle out what the presenter meant. It's better for them to interrupt by asking a question than remain confused or focused on that point for the rest of the presentation. Furthermore, if one person asks a question, chances are other people in the audience have the same question but are not comfortable or confident enough to ask.

It's also important to choose the right material for an oral presentation; some topics are more amenable to communication in this format. Analytic concepts and findings are easier to communicate in oral presentations than equations and technical details. Slides full of equations, in particular, will put an audience to sleep. Instead, discuss the implications and assumptions of the equations rather than the formulae themselves, and point the audience to a technical paper where they can find the detailed mathematics.

When planning a presentation, think about the concepts you would like the audience to take away. One or two major concepts are all most audience members will really absorb over the course of a (typical) 45-minute presentation, so consider carefully how to emphasize and support those points.

Compared to printed matter, communication during an oral presentation takes place over more "channels." For example, body language, facial expressions, tone, and gestures cannot be conveyed through printed matter. Using these non-verbal channels to reinforce your message is extremely important; your presentation should convey confidence and a mastery of the material.

Most speakers get nervous before a presentation; I know I do! Luckily, there are some simple things you can do to manage your anxiety. Practicing beforehand on an audience of friends and colleagues will help you feel more comfortable. I wouldn't recommend going as far as memorizing your presentation or writing a script, as those tactics can make your presentation sound stilted and forced. Bring a glass of water to sip from if you get flustered and bring an outline or refer to your slides if you lose your train of thought. Once you've started speaking, nervousness tends to subside. Most importantly, give as many presentations as you can, because even though experienced speakers also get nervous, managing that feeling does get easier with practice.

Oral presentations often have time limits. It is extremely important not to run over the allotted time because doing so is disrespectful to your audience and they will start to lose focus. One exception to this rule is when you are briefing a decision maker who is still actively engaging with the presentation – obviously, if they are still interested, you should proceed.

Don't overlook the use of an oral presentation as an advertisement for yourself, your work, and analytics in general. Standing up in front of an audience and defending your work, when done well, results in the audience having more trust in the findings. It's good for your reputation and can help you attract more work. At the same time, remember to follow the principle of humility.

One of the many recommendations Tufte makes in his chapter "The Cognitive Style of Powerpoint" (Tufte 2006)[5] is to write a paper handout to use as a leave-behind for the audience. The handout should be written in full sentences, like a technical paper, with data-rich graphics. Give it to the audience before the start of the presentation so they can take notes on it. Even if the audience spends some time during the presentation reading the handout, their reading your work instead of listening to the presentation is not really so bad, is it?

---

[5]These 30 pages in Tufte's book are some of the most important you will read as an analyst. They contain so much wisdom and experience; I cannot recommend them strongly enough.

11.3.3   Elevator Speeches

An elevator speech is a short (thirty to sixty seconds) description of the project you are working on, emphasizing its importance and the reasons it is interesting. An elevator speech should be given without reliance on technical language or data visualizations – it's called an elevator speech because it's the sort of explanation you would give if you only had an elevator ride's worth of time to describe your work to someone you just met.

Here's an example of an elevator speech for one of the examples included at the end of Chapter 10:

> I'm working on an analysis for the CEO to determine whether we should purchase or lease a new office building. There are a lot of variables that influence which alternative the CEO will prefer, including the company's expected growth rate, the potential rental demand for the retail shops that face the street, and whether purchasing the building will leave us with enough flexibility if our requirements change. Choosing the right alternative can make a difference of millions of dollars to our bottom line, which, given the current state of our business, could be the deciding factor between a profit and loss for the entire year.

Notice how the complexity and the importance of the project are both conveyed.

Having an elevator speech prepared will allow you to take advantage of chance encounters, which can bring your work to the attention of the leadership in your current organization or to future employers. You never know when you're going to run into someone who can influence your career, so it's important to be prepared for these moments when they arrive. To this end, you should have an elevator speech ready for all the projects you are working on, regardless of whether the project has been completed or if it is in progress. It's also important to rehearse your elevator speech, so in the pressure of the moment you don't leave out something important. Practice giving your speech to your friends and coworkers and ask them for feedback.

11.3.4   Web Pages

As the Internet becomes a more and more essential part of our lives and work, web pages have increasingly become integrated into analytic communication. Web pages are used to convey technical ideas to the general public, to communicate amongst the members of a technical team, and to announce new analytic discoveries. Compared to other communicative media, web pages have some characteristics of printed media, some of oral presentations, and some all their own.

Hyperlinks are a powerful tool that provide limited interactivity; in essence, when used intelligently they allow the audience to tune the communication to their own level of technical sophistication. Technical terms and concepts can be used without going into great detail on their definitions if a link is provided to another website where the term is defined. Similarly, technically detailed discussions can also be located off of the main page. This saves less technically savvy audience members from having to read technical discussions that would confuse or bore them. More savvy audience members do not have to read a discussion of concepts they already understand, while if they are interested in technical details they have a place to find them.

Interaction between the audience and the analyst is possible in some limited ways using websites, including message boards for public questions and debates or emails for private discussions. These media don't provide all of the interactivity available in an oral presentation, but they are a step beyond what is available in printed matter.

As mentioned in the introduction to this chapter, interactive graphics are an entirely new form of analytic communication that is made possible by the web. I won't cover that topic here, but Myatt and Johnson (2011) is a good place to start learning practical techniques, while also picking up some useful theory.

An advantage of using the web for analytic communication is that the author can update a web page as progress is made. Contrast this with the lengthy time between editions of textbooks or the slowness with which journal articles are published. Oral presentations can be put together as quickly as web pages, but they can't reach a large audience as quickly.[6]

The barrier to entry for creating websites is low, and getting lower by the day. At this point, there is barely any knowledge of coding required to create a website and hosting services are inexpensive for low volume sites. Anyone can post something on the web, which has positive and negative aspects; new ideas can spread very quickly, but they aren't vetted by any subject matter experts.

It is important for every analyst to have at least a minimal presence on the web. Just a home page stating contact information and the subject matter of your work makes it much easier for people to find you. That, of course, is the first step in communication!

People surfing the web have a much larger degree of control over the manner in which they receive information than with other methods of analytic communication. For example, people use different browsers, configure their browsers in different ways, have different resolutions on their screens, and may even be reading your website on their smart phone. There are standards for how web browsers are supposed to display pages that are coded correctly, but many browsers don't do this properly and many software programs used to build web pages don't follow current standards. All this is to say that when you are designing a website, try to make it resilient to the different ways peo-

---

[6]I expect streaming video websites will soon prove this statement false, if they haven't already.

ple will read it. Test it with a few of your friends to make sure it displays correctly.

I will close with a quick aside on printed matter that is downloaded from the web. This medium can be tricky because some readers may view your document on the screen and some may print it out to read. At the time of this writing, most high volume printers are greyscale, with the reader needing to make special effort to print on a color printer. Many readers will just print a document on a greyscale printer, and any color data visualizations will be illegible at best, and in other cases simply misleading. Hence, the options in this format are severely limited, and can be quite challenging. Redundant coding, a technical term for the practice of using two signifiers, like color and line thickness, to identify the same trait, can be useful in this situation (Ware 2013). If redundant coding is used and the colors in your document fail to print properly, the reader will still be able to use the second signifier to draw distinctions. When marking locations on a map, using points of both different shapes and colors to signify the type of location is another example of this practice.

11.4   Audience

Analysts present their work to a wide variety of audiences, ranging from other specialists in their field to the general public. Regardless of the medium being used to communicate findings, analysts must tailor the manner in which findings are presented to the audience they are addressing. Presenting material in the same way to specialists and the general public would either leave the specialists frustrated at the lack of detail or the general public confused by too much detail, jargon, and too little context.

Thinking about the perspective of the audience and your goals for communicating with them will help you tune your material appropriately. A simple categorization of audiences is helpful to generalize about these issues. Let's divide audiences into three common varieties: technical specialists, technical generalists, and the general public. As usual, this categorization doesn't have perfect mutual exclusivity or collective exhaustion, but it's a good starting point.

One piece of advice applies to all of the audience categories: don't be afraid to tell the audience something they already know, because people take a lot of pleasure in hearing or reading arguments that they agree with. This is particularly true at the start of your presentation, where you establish the context for your findings. After you set the scene, you can then discuss more controversial conclusions.

11.4.1 The Technical Specialists

By <u>technical specialists</u>, I am referring to anyone that knows the topic of your work as well or better than you or someone who, by the end of your presentation, must reach that level of understanding. Technical specialists include your coworkers, supervisor, and competitors. However, don't include technical specialists from other specializations in this category; those people should be treated as technical generalists, discussed in the next section.

Technical specialists use a variety of media to communicate with each other. Many fields have peer-reviewed journals in which technical ideas are presented, universities arrange presentations for the exchange of technical ideas, technologically oriented businesses write working papers, and all analysts use field conferences to present their work to colleagues. Presentations in each of these media have some common themes because they are all made to technical specialists.

In my experience, it is easier to communicate analytic ideas to other technical specialists than to any of the other groups because the audience and the presenter have similar backgrounds. In particular, people in the same specialty have a common history of solved problems from earlier in their education. This means that a lot of the explanatory work has already been done – basically there is a short distance between the ways that the presenter and the audience think about the work. Drawing analogies that depend on the understanding of simple toy problems that were taught in specialized classes can be an effective technique with this audience.

Additionally, when addressing other technical specialists, less effort needs to be spent explaining techniques (with the exception of those that are novel). For example, if your field has specialized visualization techniques that are commonly used, you won't need to thoroughly explain them because the audience will have seen similar communication tools before. Similarly, field jargon is appropriate for this audience. That doesn't mean that you should fill your presentation with jargon and acronyms; clarity is still a guiding principle, but technical terms do allow conversational precision and efficiency.

Even when presenting to technical specialists, it is important to spend some time in the introduction and conclusion explaining how your work fits into the larger context. This demonstrates your understanding of the state of the field and how you and the other analysts in the audience fit into it. Unlike with most audiences where you'll want to focus on communicating one or two ideas, with other technical specialists you can stretch to three or four ideas, unless you have very limited presentation time or page space to work with.

11.4.2 The Technical Generalists

<u>Technical generalists</u> are people who have some detailed knowledge of your field, but have specialized in some other subject or are not up-to-date on the latest happenings in the field because they are filling some other role

in your organization (such as fund-raising, project management, or business development). They don't need to know all of the nitty-gritty details of your work, but they do need to know how the work was carried out in some detail and they'll be interested to know what sets it apart from previous efforts. Examples of technical generalists include the leadership level of a business and technical specialists from other fields.

Media in which analysts communicate with technical generalists include broad journals like *Science* and *Nature*, specialized magazines like *Analytics* or *OR/MS Today*, grant applications and other requests for funding, and department-wide colloquia in academia. Communicating with a technical generalist audience is of increasing importance due to the modern emphasis on interdisciplinary work.

When using analogies to help a technical generalist audience understand a new concept, it's a good idea to rely on toy problems from an introductory level only. That's about the only background that everyone in the audience has in common. When using these analogies, present a short review of the original toy problem as well, since for much of your audience it has been a long time since those introductory classes.

If any jargon or technical terms are to be used, they should be clearly defined first. Try to keep these to a minimum because you don't want to waste time and effort on definitions; a good rule of thumb is that it's only worth it to introduce a technical term or acronym if you are going to use it three or more times throughout your presentation. If not, explain the concept without the specialized vocabulary and save the time to communicate ideas.

Similarly, you may need to repackage visualizations that are common in your field for a technical generalist audience, particularly if the visualization doesn't follows standard analytic conventions. There's nothing wrong with doing this – many visualization conventions are passed down from arbitrary choices made decades ago, and make little sense today.

In my experience, technical generalists are the most challenging audience to communicate with. There can be little common background and some members of the audience will find your findings so far removed from their own work that it is hard to keep them focused. Similarly, different fields approach problems in varied ways, so they may find fault with techniques that are common to your field. For these reasons, a technical generalist audience requires a tough balancing act between detail and context.

11.4.3 The General Public

One of the great pleasures of being an analyst is the opportunity to present your technical work to people who have no background in your field. These opportunities are a chance to describe how your findings fit into the big picture, why your work makes you want to get out of bed in the morning, and why they should care about it too!

Communicating with the general public occurs over a variety of media.

Imagine, for example, a researcher writing about her work for her university's public website, a scientist giving an interview to a news organization, or even a financial analyst giving a speech about a company's prospects to shareholders.

In fields with a long history (such as the sciences), very little time should be spent discussing your own work. The vast majority of the presentation should focus on the larger context, explaining the historical development of the field with a little bit of your own work sprinkled in. For fields in which historical development is less important (such as the financial analyst mentioned above), most of the time should be spent on the interpretation of the findings. Don't leave that interpretation unsaid or only understood implicitly.

Jargon and acronyms should be minimized and any visualizations should be explained in excruciating detail. In a public talk, the rate at which information is presented should be fairly slow; don't try to flash a new slide up on the screen every thirty seconds because you'll lose the audience.

When explaining technical matters, illustrate the concepts with analogies to everyday life and common experiences. These examples, while rarely perfect matches, will stick with the audience long after your presentation is over. When I was a graduate student, I was explaining a piece of technical work to a general audience, and was struggling to describe a geometrically complex disc shape. My graduate advisor, Professor Leo Blitz, who had far more experience speaking to general audiences, was sitting at my side. He saw I was struggling and summarized my long-winded explanation with a sentence: "It's like a vinyl record that's been left out in the sun." Of course, that description was the audience's takeaway from the presentation, but if the audience had been left with my more long-winded description they would have retained nothing. It wasn't an exact fit to the science, but it was pretty close, it was understood by everyone, it was pithy, it brought to mind a mental picture, and it was memorable.

The best advice I can give you is to attend other specialists' presentations for a general audience and to read their articles. Do this both for specialists in your own field, so you can see how they present the concepts you work with everyday, and for specialists in other fields, so you can see what techniques you find effective as an audience member. Then, when the time comes for you to give your own presentation for the general public, practice with friends that don't specialize in your field. It'll quickly become apparent to you what works and what doesn't.

11.5 Tips for Data Visualization

After these general discussions of analytic communication objectives, media, and audiences, I would be remiss not to mention some tips for building data visualizations. This is a very active and important subfield, of which I am only going to touch on a few select topics. For a comprehensive introduction, the best starting point is Tufte (2001). His work certainly has influenced all the advice I provide.

First, following Iliinsky and Steele (2011), I will distinguish between exploratory and explanatory data visualizations. An <u>exploratory visualization</u> is one that is used by the analyst to better understand the data. On the other hand, <u>explanatory visualizations</u> are used by an analyst to explain to an audience something about the data. These two types of data visualizations have very different functions, and therefore different forms.

Exploratory data visualizations are generally rich with detail, much of it extraneous, because the analyst is still trying to figure out what the key findings are. They are, in essence, tools for *doing* analysis. These visualizations' usage as investigatory tools and the fact that they are not intended to be shared beyond the analyst (and possibly a few close team members or colleagues) means that they are not polished, because the audience is presumably willing to expend a lot of effort to understand the data. That's not to say that exploratory visualizations should be purposely obtuse but that they can err on the side of including more information because it isn't yet clear what is important.

Instead of being a tool for doing analysis, explanatory data visualizations are tools for *communicating* analytic findings. These visualizations are a product of analysis, rather than an intermediate step. Because they are intended to be viewed and understood by a large audience, they must be polished and crystal clear.[7] They should also be properly tuned for the type of viewing audience. The remainder of this section will concentrate on explanatory, rather than exploratory, visualizations.

The first thing to know about data visualization is that making good plots takes a really long time. When constructing an explanatory visualization, you will need to make a large number of small decisions. For example, you'll need to choose labels, colors, and tickmarks. Don't just settle for the software defaults; in many cases, they aren't the best choice for a specific plot. You should take care with every pixel, examine the plot from every angle. This will require you to get into the nitty gritty details of whatever software program you are using to create the visualization. When in doubt, let the clarity principle be your guide.

Clarity doesn't require emphasizing simplicity to the exclusion of everything else. Consider, for example, the concept of redundant coding, introduced in Section 11.3.4. One reason for using redundant coding is that one of your signifiers may fail, so the other functions as a failsafe. A second is that it allows the audience members to search a visualization using their choice of two different criteria, whichever is easiest for them. In plots that feature many lines, it can be better to keep things simple and vary only one feature, but

---

[7]It is certainly possible for a visualization to initially be created for exploratory purposes and then later to be cleaned up for explanatory use.

in simple plots you can use redundant coding to maximally differentiate your curves.

Color is a powerful tool for differentiation. Colors that represent different curves should be as easy as possible to distinguish. For example, if you have two lines on a plot, don't draw one in dark blue and the other in black. I've found that bright red and dark black are quite easy to tell apart. Avoid plots that directly contrast red and green; 7% of the male population (and 0.4% of females) have trouble distinguishing between those two colors, which probably includes someone in your audience (Montgomery 1995). A less common form of color blindness causes difficulty distinguishing between blue and yellow.

As a trained astronomer it's hard for me to admit this, but most people don't internalize a color's hue as an ordered quantity. In other words, a portion of your audience may know the order of the spectrum is ROYGBIV,[8] but they don't internalize hue as an ordered series (e.g., red is less than blue) (Tufte 1990). However, brightness and saturation, two additional properties that together with hue make up color, are ordered (Iliinsky and Steele 2011). It's best to use hue for categorical relationships and brightness or saturation for ordinal relationships.

Adding a layer of annotation directly on top of a visualization ensures that the audience will not miss the important ideas you want them to take away. Label boxes with arrows pointing to interesting regions in the data are a powerful tool for attracting the audience's attention. They can be effective entry points for drawing in the audience to the rest of your material.

If your data visualization is going to appear in multiple media, you will need to make different versions of it. For example, if you have a simple visualization of 5 points with error bars on x-y axes, make a version of it with thin lines and error bars for a paper, but also make a version with larger symbols, thicker lines, and large font axis labels for use in slides.

If your oral presentation is going to be accompanied by projected slides, take special care with the visualizations you include. If the visualization is going to be in an oral presentation, some methods of differentiating lines can show up poorly in low resolution media, such as thickness or dots/dashes. Also, yellow tends to show up poorly on a projection screen. Symbols and fonts (including axis labels) need to be much larger than on a paper visualization, and lines should be thicker. Note that a well-designed visualization on a slide looks aesthetically terrible when printed on a piece of paper – don't let that dissuade you.

When I'm struggling to effectively create a visualization for a complicated data set, I usually flip through other visualizations to look for inspiration. Often I'll run into someone who has built an elegant visualization for a similarly structured data set. Tufte's books are a great place to start when browsing in this way, but I've found inspiration in some surprising places as well because

---

[8]That's red-orange-yellow-green-blue-indigo-violet, from longer wavelengths to shorter wavelengths.

many people in other fields are doing amazing things with data visualization. For example, browse Chris Ware's graphic novels (Ware 2000), Randall Munroe's webcomics (www.xkcd.com), Kate Ascher's book on how cities function (Ascher 2005), or Mark Danielewski's novels (Danielewski 2000).

Finally, here is one last piece of advice regarding data visualizations: because they are powerful means of communicating large amounts of information, it is best not to assume that your plot will stay within one degree of separation of you. If you email a copy of it to your boss, he's likely to email it to his boss without checking with you first. With that in mind, whenever you make a plot for someone else's consumption, make it look good. Once it leaves your desktop in any form, it is out of your control. Even if you think the plot is only for a very small audience, it may eventually wind up on the desk of someone very important, and when given that kind of a lucky break you should take advantage of it and make a good impression.

11.6 Exercises

1. Modify Figure 3.1 for use in an oral presentation, projected onto a large screen. What changes did you make?

2. Add an annotation layer to one of the visualizations you've constructed in the course of completing this textbook. Make one version for technical specialists, and a second for the general public.

3. Write an elevator speech for a past project on which you have worked.

12

What to Do Next

This book is a practical introduction to analytics, but analysts should grow and learn new things throughout their careers. What should you do to continue your growth as an analyst? This final chapter is a list of actions you can take to improve your technical skills and analytical abilities.

- **Grapple with real problems**
  The surest way to improve your analytic skills is to solve real problems. Ideally, you should do this outside of the classroom; there's nothing like the real world and real data to clarify what works and what doesn't. Always be on the lookout for interesting problems and decisions with which you can be involved. If you can't find anything, work with case studies built around real situations. Learn which styles and approaches work for you.

- **Apply analytics to your own decision making**
  In Chapter 9, I described several techniques that you can apply to your own decision making. Start to use them in your own life; this will help you appreciate the perspective of the decision maker in addition to that of the analyst. Don't just use these techniques on the important decisions in your life – also use the them for decisions that are interesting, such as decisions in which your payoffs are tied to the decisions of other people or those in which the uncertainty is large.

- **Learn a little bit about a lot of techniques**
  Analytics is a rich and broad field; there are many other textbooks, professional journals, and web pages focusing on its many subfields. Expose yourself to as many of these topics as possible, not because you should master them all, but because you need to be familiar with the variety of techniques available. Then, when you see a problem that is amenable to one of those techniques, you'll recognize it and you can learn how to use it in detail.

- **Make analytic friends**
  One of the best ways to learn about new analytic techniques is from other people who have successfully applied them. It really helps to have a group of friends working in analytics who can share experiences and skills. Such a group is also useful to bounce ideas off of when you run into a problem. The strongest groups are interdisciplinary, in that they involve people from different backgrounds with varied training. Consider setting up a regular

lunch or dinner to meet and chat. An email list to share stories and ask for advice also helps.

- **Join a professional society**
  There are many professional societies relevant to analytics. Joining a professional society is an important step in becoming part of the larger analytic community and establishing yourself as an analyst. A society will help you stay up-to-date on current developments and trends in the field. Some examples of relevant societies are the Institute for Operations Research and Management Science (INFORMS, www.informs.org), which has started an analytics section to promote data-driven analytics and fact-based decision making, the Society for Risk Analysis (www.sra.org), and the American Association for the Advancement of Science (www.aaas.org).

- **Attend a conference**
  Another way to hear about how analytics is being applied is to listen to presentations from other practitioners. Conferences are a great place to learn new techniques, meet other practitioners, and form working collaborations. They are also great places for networking and job hunting. INFORMS is currently holding a few relevant conferences: in the United States, there is the long-established annual meeting, which has a primarily academic bent, and the newer annual conference on business analytics and operations research, which is more focused on practice. INFORMS also runs a regular international conference.

- **Practice leading**
  The best way to practice being a leader is to actually lead a project. It's important to feel that responsibility on your shoulders; it's not something easily replicable in a classroom environment.

  Often, analysts will be in the position of bringing analytics to an organization that doesn't currently use it. This is actually a common leadership situation, also known as change management. There is a cottage industry of books full of leadership advice on how to best manage these situations. Humphrey (1997) has some good advice on being a technical leader and I also enjoyed Belsky (2010), which is focused more on general advice about leadership and less on analytical ideas. Part of being a leader is knowing how to operate inside of an organization; the best book I've ever read on organizations, particularly large ones, is *Bureaucracy* (Wilson 1989).

- **Improve your programming skills**
  As I mentioned in the preface, this book does not require you to be skilled in programing, but in the long run if you want to pursue analytics you are going to need to bite that bullet. Fortunately, you can actually teach yourself how to program by following along with an introductory textbook or with materials on the web. Programing classes can help you learn the theory behind algorithms, which isn't strictly necessary for analytics, but certainly is helpful.

If you already know how to program, try to push yourself outside your comfort zone with the tasks you accomplish by coding. Pick up a new language, write code on a different operating system, build programs with web interfaces, or learn how to use version control software. The programing world is very fast paced, and it takes effort not be left behind by evolving techniques.

- **Teach or tutor less experienced analysts**
  Teaching and tutoring serves a number of purposes. First, I've always found that explaining technical material to someone else helps me understand the material better. It organizes my thoughts and forces me to study up and eliminate gaps in my knowledge so I can answer questions. Second, as I discussed in Chapter 11, the ability to talk about a technical subject to people who are not technically trained is an essential skill for an analyst. Finally, each of us should be an advocate for others to understand and use analytics – there are more than enough problems amenable to analysis to go around, and we need all the help we can get. Even if the person you teach doesn't become an analyst, you can help them recognize analytic problems so they will know when to ask for help from an analyst.

- **Get certified**
  There are a number of professional certifications available that add to your credibility as an analyst and sharpen the skills you need to perform analytic work. In Chapter 10, I mentioned the project management certification – this is an excellent place to start. Six Sigma^{TM} is another certification available for those interested in designing analytic processes. For some financial analysts, an accounting certification could be worthwhile.

  At the time of writing, the INFORMS Analytics section is beginning an analytics certification curriculum and examination. There are also a growing number of graduate Master's programs focusing on analytics.

  I find analytics to be a tremendously exciting field, and I hope that you will too. I really feel that the field is poised to take off, and I'm glad that you will be a part of it. All that's left for me to do is to wish you the best of luck in your endeavors, analytic and otherwise.

Bibliography

Amante, C. and Eakins, B. (2008). ETOPO1 1 arc-minute global relief model: Procedures, data sources and analysis. Technical report, National Geophysical Data Center, NESDIS, NOAA, U.S. Department of Commerce.

Apgar, V. (1953). A proposal for a new method of evaluation of the newborn infant. *Current Research in Anasthesia and Analgesia*, **32**(4), 260–7.

Ascher, K. (2005). *The Works: Anatomy of a City*. Penguin Books, New York, NY.

Augustine, N. (1997). *Augustine's Laws*. American Institute of Aeronautics and Astronautics, Reston, VA, 6th edition.

Bagchi, S. and Mitra, S. (2001). The nonuniform discrete fourier transform. In F. Marvasti, editor, *Nonuniform Sampling: Theory and Practice*. Kluwer Academic/Plenum Publishers, New York, NY.

Bedford, T. and Cooke, R. (2001). *Probabilistic Risk Analysis*. Cambridge University Press, Cambridge, UK.

Belsky, S. (2010). *Making Ideas Happen*. Penguin Books, New York, NY.

Berger, M. and Colella, P. (1989). Local adaptive mesh refinement for shock hydrodynamics. *Journal of Computational Physics*, **82**, 64–84.

Bevington, P. and Robinson, D. (2003). *Data Reduction and Error Analysis for the Physical Sciences*. McGraw Hill, New York, NY, 3rd edition.

Bretherton, F. P. and Singley, P. (1994). Metadata: A user's view. In *Proceedings of the International Conference on Very Large Data Bases*, pages 1091–1094.

Brooks, F. J. (1995). *The Mythical Man Month*. Addison-Wesley, Reading, MA.

Brown, L., Cai, T., and DasGupta, A. (2001). Interval estimation for a binomial proportion. *Statistical Science*, **16**(2), 101–117.

Bruce, A. and Gao, H. (1996). *Applied Wavelet Analysis with S-PLUS*. Springer-Verlag, New York, NY.

Clemen, R. and Reilly, T. (2001). *Making Hard Decisions*. Duxbury, Pacific Grove, CA.

Cohen, D., Lindvall, M., and Costa, P. (2004). An introduction to agile methods. *Advances in Computers*, **62**, 1–66.

Cooley, J. and Tukey, J. (1965). An algorithm for the machine calculation of complex Fourier series. *Mathematics of Computation*, **19**, 297.

Danielewski, M. (2000). *House of Leaves*. Pantheon, New York, NY.

Davenport, T. and Harris, J. (2007). *Competing on Analytics*. Harvard Business School Press, Cambridge, MA.

Davenport, T., Harris, J., and Morison, R. (2010). *Analytics at Work*. Harvard Business Press, Cambridge, MA.

Deitel, H. and Deitel, P. (2001). *C++: How to Program*. Prentice Hall, Upper Saddle River, NJ.

Emerson, J. and Hoaglin, D. (1983). Stem-and-leaf displays. In D. Hoaglin, F. Mosteller, and J. Tukey, editors, *Understanding Robust and Exploratory Data Analysis*, pages 7–32. John Wiley & Sons, New York, NY.

Falkowski, M., Smith, A., Hudak, A., Gessler, P., Vierling, L., and Crookston, N. (2006). Automated estimation of individual conifer tree height and crown diameter via two-dimensional spatial wavelet analysis of lidar data. *Canadian Journal of Remote Sensing*, **32**(2), 153–161.

Feynman, R. (1997). *Surely You're Joking, Mr. Feynman!* W.W. Norton & Co., New York, NY.

Frankfort-Nachmias, C. and Nachmias, D. (1992). *Research Methods in the Social Sciences*. St. Martin's Press, New York, NY, 4th edition.

Goldratt, E. (1997). *Critical Chain*. The North River Press, Great Barrington, MA.

Hammond, J., Keeney, R., and Raiffa, H. (1999). *Smart Choices*. Broadway Books, New York, NY.

Heiles, C. (2005). Discreetly fine times with discrete Fourier transforms. Notes at astro.berkeley.edu/~heiles/handouts.

Heiles, C. (2008). Least-squares and chi-square for the budding aficionado. Notes at astro.berkeley.edu/~heiles/handouts.

Hora, S. (2007). Eliciting probabilities from experts. In W. Edwards, R. J. Miles, and D. von Winterfeldt, editors, *Advances in Decision Analysis*, chapter 3. Cambridge University Press, Cambridge, UK.

Humphrey, W. (1997). *Managing Technical People.* Addison Wesley, Reading, MA.

Hwang, J., Cho, S., Moon, J., and Lee, J. (2005). Nonuniform DFT based on nonequispaced sampling. In *Proceedings of the 5th WSEAS Int. Conf. on Signal, Speech and Image Processing*, pages 11–16.

Iliinsky, N. and Steele, J. (2011). *Designing Data Visualizations.* O'Reilly, Sebastopol, CA.

Inmon, W. and Nesavich, A. (2008). *Tapping Into Unstructured Data.* Prentice Hall, Upper Saddle River, NJ.

Jones, P. D. and Moberg, A. (2003). Hemispheric and largescale surface air temperature variations: An extensive revision and an update to 2001. *Journal of Climate*, **16**, 206–223.

Kamath, C. (2009). *Scientific Data Mining: A Practical Perspective.* SIAM, Philadelphia, PA.

Keeney, R. (1993). *Value-Focused Thinking.* Harvard University Press, Cambridge, MA.

Keeney, R. (2007). Developing objectives and attributes. In W. Edwards, R. J. Miles, and D. von Winterfeldt, editors, *Advances in Decision Analysis.* Cambridge University Press, Cambridge, UK.

Keeney, R. and Gregory, R. (2005). Selecting attributes to measure the achievement of objectives. *Operations Research*, **53**(1), 1–11.

Keeney, R. and Raiffa, H. (1976). *Decisions with Multiple Objectives.* Cambridge University Press, Cambridge, UK.

Kendall, M. (1938). A new measure of rank correlation. *Biometrika*, **30**(1-2), 81–93.

Kent, S. (1964). Words of estimative probability. *Studies in Intelligence*, **8**, 49–65.

Knight, F. (1921). *Risk, Uncertainty, and Profit.* Riverside Press, Cambridge, MA.

Knuth, D. (1998). *The Art of Computer Programming*, volume 3. Addison-Wesley, Reading, MA, 2nd edition.

Kopka, H. and Daly, P. (1999). *A Guide to LATEX.* Addison Wesley, Harlow, England, 3rd edition.

Kreyszig, E. (1993). *Advanced Engineering Mathematics.* John Wiley & Sons, New York, NY.

Krieg, M. (2001). A tutorial on bayesian belief networks. Technical report, Australian Department of Defence, http://www.dtic.mil/cgi-bin/GetTRDoc?AD=ADA401153.

Levine, E., Blitz, L., and Heiles, C. (2006). The spiral structure of the outer Milky Way in hydrogen. *Science*, **312**(5781), 1773–7.

Lewis, M. (2010). *The Big Short*. W.W. Norton & Co., New York, NY.

Lomb, N. R. (1976). Least-squares frequency analysis of unequally spaced data. *Astrophysics and Space Science*, **39**, 447–462.

Lord, F. (1953). On the statistical treatment of football numbers. *American Psychologist*, **8**(12), 750–751.

Lustig, I., Dietrich, B., Johnson, C., and Dziekan, C. (2010). The analytics journey. *Analytics Magazine*, (November/December).

Mandel, J. (1964). *The Statistical Analysis of Experimental Data*. Dover Publications, New York, NY.

Markel, M. (2012). *Technical Communication*. Bedford/St. Martin's, Boston, MA, 10th edition.

Marvasti, F. (2001). Random topics in nonuniform sampling. In F. Marvasti, editor, *Nonuniform Sampling: Theory and Practice*. Kluwer Academic/Plenum Publishers, New York, NY.

Miller, R. (1963). *Schedule, Cost, and Profit Control with PERT*. McGraw-Hill, New York, NY.

Montgomery, G. (1995). Breaking the code of color. In *Seeing, Hearing, and Smelling the World*. Howard Hughes Medical Institute, Chevy Chase, MD.

Mosteller, F. and Tukey, J. (1977). *Data Analysis and Regression*. Addison Wesley, Reading, MA.

Myatt, G. and Johnson, W. (2011). *Making Sense of Data III: A Practical Guide to Designing Interactive Data Visualizations*. Wiley, Hoboken, NJ.

Oppenheim, A., Schafer, R., and Buck, J. (1999). *Discrete-Time Signal Processing*. Prentice Hall, Upper Saddle River, NJ, 2nd edition.

Parnell, G. (2007). Value-focused thinking using multiple objective decision analysis. In A. Loerch and L. Rainey, editors, *Methods for Conducting Military Operational Analysis: Best Practices in Use Throughout the Department of Defense*. Military Operations Research Society, Alexandria, VA.

Peterson, T. and Vose, R. (1997). An overview of the global historical climatology network temperature database. *Bulletin of the American Meteorological Society*, **78**, 2837–2849.

Pourret, O., Naim, P., and Marcot, B. (2008). *Bayesian Networks: A Practical Guide to Applications*. Wiley, Hoboken, NJ.

Press, W. H., Teukolsky, S. A., Vetterling, W. T., and Flannery, B. P. (2007). *Numerical Recipes*. Cambridge University Press, Cambridge, UK, 3rd edition.

Raiffa, H. (2007). Decision analysis: A personal account of how it got started and evolved. In W. Edwards, R. J. Miles, and D. von Winterfeldt, editors, *Advances in Decision Analysis*, chapter 4. Cambridge University Press, Cambridge, UK.

Raiffa, H., Richardson, J., and Metcalfe, D. (2002). *Negotiation Analysis*. Belknap Press, Cambridge, MA.

Reynolds, G. (2008). *Presentation Zen*. New Riders, Berkeley, CA.

Rosolowsky, E. (2005). Wavelets. Notes can be downloaded at astron.berkeley.edu/~heiles/handouts/wavelet.ps.

SAS (2008). Eight levels of analytics. *SASCOM Magazine*, (Fourth Quarter).

Savitzky, A. and Golay, M. (1964). Smoothing and differentiation of data by simplified least squares procedures. *Analytical Chemistry*, **36**, 1627.

Scargle, J. D. (1982). Studies in astronomical time series analysis. II - Statistical aspects of spectral analysis of unevenly spaced data. *Astrophysical Journal*, **263**, 835–853.

Schachter, R. (2007). Model building with belief networks and influence diagrams. In W. Edwards, R. J. Miles, and D. von Winterfeldt, editors, *Advances in Decision Analysis*. Cambridge University Press, Cambridge, UK.

Silver, N. (2012). *The Signal and the Noise*. Penguin Press, New York, NY.

Sivia, D. and Skilling, J. (2006). *Data Analysis: A Bayesian Tutorial*. Oxford University Press, Oxford, UK, 2nd edition.

Sorkin, A. (2009). *Too Big to Fail*. Penguin Books, New York, NY.

Spearman, C. (1904). The proof and measurement of association between two things. *American Journal of Psychology*, **15**, 72–101.

Stevens, S. (1946). On the theory of scales of measurement. *Science*, **103**, 677–680.

Stevens, S. (1951). *Handbook of experimental psychology*, chapter Mathematics, measurement and psychophysics, pages 1–49. Wiley, Hoboken, NJ.

Stewart, T. (1996). Robustness of additive value function methods in MCDM. *Journal of Multicriteria Decision Analysis*, **5**(4), 301–309.

Strunk, W. and White, E. (1959). *The Elements of Style*. Macmillan, New York, NY, 1st edition.

Taleb, N. N. (2007). *The Black Swan: The Impact of the Highly Improbable*. Random House, New York, NY.

Taylor, J. (1997). *An Introduction to Error Analysis*. University Science Books, Sausalito, CA, 2nd edition.

Torrence, C. and Compo, G. (1998). A practical guide to wavelet analysis. *Bull. of the Amer. Meteorological Soc.*, **79**, 61.

Tufte, E. (1990). *Envisioning Information*. Graphics Press LLC, Cheshire, CT.

Tufte, E. (2001). *The Visual Display of Quantitative Information*. Graphics Press LLC, Cheshire, CT, 2nd edition.

Tufte, E. (2006). *Beautiful Evidence*. Graphics Press LLC, Cheshire, CT.

Tukey, J. (1977). *Exploratory Data Analysis*. Addison Wesley.

Velleman, P. and Wilkinson, L. (1993). Nominal, ordinal, interval, and ratio typologies are misleading. *The American Statistician*, **47**(1), 65–72.

Virine, L. and Trumper, M. (2008). *Project Decisions: The Art and Science*. Management Concepts, Vienna, VA.

Von Neumann, J. and Morgenstern, O. (1944). *Theory of Games and Economic Behavior*. John Wiley & Sons, New York, NY.

Ware, C. (2000). *Jimmy Corrigan: The Smartest Kid on Earth*. Pantheon, New York, NY.

Ware, C. (2013). *Information Visualization: Perception for Design*. Morgan Kauffman, New York, NY, 3rd edition.

Weiss, S., Indurkhya, N., and Zhang, T. (2010). *Fundamentals of Predictive Text Mining*. Springer, London, UK.

Wilson, E. (1952). *An Introduction to Scientific Research*. Dover Publications, New York, NY.

Wilson, J. (1989). *Bureaucracy*. Basic Books, New York, NY.

Index

1-dimensional data sets, **75**, 75–123
  contrast, increasing, 97–99
  examples, 79–82
  filtering, 114–115
  fitting, 88–97
  frequency analysis of, 99–113
  interpolation, 82–85
  smoothing, 85–88
  spacing of, 76–79
  wavelet analysis of, 115–121
  see also related 1-dimensional data sets
2-dimensional data sets, **149**, 149–172
  contour plots, 156–159
  correlation coefficients, 166–168
  correlation functions, 170
  dimensionality and, 150
  examples, 151–152
  formalism, 149
  frequency analysis, 160–164
  interpolation, 154–156
  ratios, 169
  restructuring, 153–154
  scatter plots, 166
  smoothing, 159–160
  wavelet analysis, 164–166

accuracy, **63**
age list, **31**
  fitting, 93–95, 97
  frequency distribution, 35
  histogram, 38–40
  mean, 35
  mean absolute deviation, 47
  median, 36
  mode, 37
  outliers, 35
  parent distribution, 60
  standard deviation, 45, 46
  stem and leaf diagram, 41, 43
  variance, 44
algorithms, 32
aliasing, **102**
  2-dimensional, 163
alternatives, **188**
  constraints and, 191
  decision trees and, 208
  evaluating, 192, 194, 200–206
  game theory and, 210
  identifying, 197–200
amplitude, **100**
analysts, skills of, 2–3
analytics, **1**
  applied to projects, *see* project, management
  certifications, 257
  communicating, *see* communication
  conferences, 256
  descriptive, **1**
  in groups, 7–9
  predictive, **1**
  prescriptive, **1**, 5, *see also* prescriptive decision analysis
  process of, 3–7
  professional societies, 256
  three types of, 1
angular frequency, **100**
attributes, **21**, 21–26

constructed, **23**, 23–25, 221
evaluating alternatives with, 199
measuring objectives with, 195
natural, **22**, 22–23
  age list and, 31
  height list and, 30
negotiations and, 213
proxy, **25**, 25
swing weights and, 205
audience, 248–251
  general public, 250–251
  technical generalists, **249**
  technical specialists, **249**
autocorrelation function, **147**, 170
average, *see* mean
average absolute deviation, *see* mean absolute deviation

Bayesian techniques, 178–186
  Bayes' rule, **178**, 178–181
  Bayesian statistics, **185**, 185–186
  belief networks, **181**, 181–185
  conditional probability, **178**, 180, 182
  posterior probability, **180**
  prior probability, **179**, 182
  uncertainty and, 64
bias, *see* error, systematic

causal network, **181**, 182, 184
  node, **181**
ceiling, 37
Chauvenet's criterion, **54**, 54–55
χ^2, 51
climate data set, **152**
  3-dimensional structure, 170
  contour plot, 159, 167, 168
  Pearson correlation coefficient, 168
  ratios, 169
  scatter plot, 169
collaboration, 8

collectively exhaustive, **15**
  audience categories, 248
  decision tree branches, 207
  fundamental objectives, 194
  objectives hierarchy, 194
color (in data visualization), 253
communication, 7, 237–254
  media, 242–248
    elevator speeches, 246
    oral presentations, 244–245
    printed matter, 243–244
    web pages, 246–248
  principles of, 238–242
    clarity, **239**
    humility, **242**, 245
    integrity, **241**, 241–242
    transparency, **240**, 240–241
conditional probability, **178**, 180, 182
constraints (in prescriptive decision analysis), **191**
contour plots, 156–159, 161, 167, 168
contrast, increasing, 97–99
correlation, **129**, 129–138
  2-dimensional data sets and, 166
  product-moment coefficients, **134**, 134–137
  rank coefficients, **137**, 137–138
  scatter plots, **131**, 131–134
correlation functions, 143–147
  2-dimensional, 170
  autocorrelation, 147
  cross-correlation, 143–147
critical path, **219**, 228, 230, 231, 236
cross-correlation function, **143**, 143–147, 170

data set, 13, **26**
data visualization, **237**, 251–254
  explanatory, **252**, 252–254
  exploratory, **252**

deciles, 36
decision, **188**
  basic structure of, 188
  environment, **189**, 189–192, 196, 198
  informing a, 188
  new car example
    attributes of, 206
    decision environment of, 191
    identifying alternatives, 199
    multi-attribute utility of, 205–206
    objectives of, 195–196
    single-attribute utilities of, 202
decision maker, **188**
  as part of the decision environment, 189
  identifying alternatives and, 198
  objectives of, 192
  utility function of, 201–205
decision trees, **206**, 206–208
direct current, **105**, 109
discrete Fourier transform (DFT), **104**, 104–109
  2-dimensional, 161–165
dispersion, **43**
distribution, **47**, 59
  bimodal, 44, 46
  frequency, *see* frequency distribution
  limiting, *see* distribution, parent
  normal, 30, 39, 45, **48**, 48–49
    problems with, 71–73
  parent, **59**, 69
  sample, **59**
  uncertainty, **64**
  see also frequency distribution
dominant strategy, 210
dominate, **34**
dynamic range, **31**

element spacing, 76–79
  circular, **78**, 86, 121
  evenly spaced, **76**
  evenly spaced with gaps, **76**, 86
  unevenly spaced, **78**, 88
elements, **13**, 26
elevation data set, **80**, 81
  cross-correlation, 145, 146
  Kendall τ correlation coefficient, 138
  median smoothing, 88, 89
  north of the equator, **127**, 129
  Pearson correlation coefficient, 136
  ratio, 140, 142
  scatter plot, 133, 135
  wavelet analysis, 121, 122
elevation map, **151**
  contour plot, 157, 158
  restructuring, 153, 154
  smoothing, 160, 161
error, **58**
  illegitimate, **58**
  random, **60**
  systematic, **61**, 69
  see also uncertainty
error bar, 64
errors of omission and commission, **178**
Euler's formula, 105
expected utility, **201**, 208
expected value, **201**, 202, 208

false negative, **180**
false positive, **179**, 180
fast Fourier transform (FFT), **109**, 162
filtering, **114**
  2-dimensional, 163
fitting, **88**, 88–97
  dependent variable, **89**
  functional form, **90**
  general linear, 95–97
  independent variable, **90**

least-squares, **90**
parameters, **90**, 91
related data sets, 142–143
straight line, 92–95
floor, 37
Fourier transform, *see* discrete Fourier transform
frequency, **99**
frequency analysis, **99**, 99–113
    2-dimensional, 160–164
    Fourier algorithms, **102**, 102–111
        discrete Fourier transform (DFT), 104–109
        fast Fourier transform (FFT), 109, 162
        Fourier expansion, **103**
        Fourier transform, **104**
        inverse discrete Fourier transform, **108**
        power spectrum, **104**, 110–111
    Lomb periodogram, 111–113
frequency distribution, 32, **33**, 47
    histogram and, 38
    mean, use in calculating, 34
    mode and, 36
    stem and leaf diagram and, 41
    variance and, 43
frequentist statistics, **185**
Full-width half-maximum (FWHM), **48**

game theory, **209**, 209–212
Gantt chart, **228**, 228–231, 233, 236
greyscale image, **151**
    discrete Fourier transform (DFT), 162, 165
    restructuring, 153
    visualization, 162, 164

height list, **30**
    Chauvenet's criterion and, 55
    frequency distribution, 49, 51–52

histogram, 38, 39, 60, 61
Kendall τ correlation coefficient, 138
mean, 35
mean absolute deviation, 47
median, 36
mode, 37
outliers, 31, 35, 37, 39, 44, 47
parent distribution, 59
Pearson correlation coefficient, 136
ratio, 139, 140
scatter plot, 132, 133
standard deviation and, 46
stem and leaf diagram, 41, 42
uncertainty in the mean, 68
variance, 44
weight list, **127**, 128
heterogenous dimensionality, **150**, 171
histogram, **38**, 38–41
    hypothesis testing and, 53
    parent distribution and, 61
    ratios and, 140
homogenous dimensionality, **150**, 171
hypothesis testing, **50**, 50–52

INFORMS, 1, 3, 256, 257
interpolation, **82**, 82–85, 154–156
    2-dimensional, 154–156
    bilinear, **155**
    Fourier transforms and, 109, 163
    linear, **83**, 156

Kendall τ rank correlation coefficient, **137**

lessons learned report, 231
levels of measurement, 14–20
lists, **29**, 29–56
    elementary analyses (mean, median, mode), 32–37
    examples, 30–31
    formalism, 29

histogram, 38–41
mean absolute deviation, 46–47
standard deviation, 45–46
stem and leaf diagrams, 41–42
variance, 43–45
Lomb periodogram, **111**, 111–113
wavelet analysis, use with, 116

margin of error, **70**
mean, **33**, 33–36
bimodal distribution and, 44
ordinal scales and, 17
mean absolute deviation, **46**, 46–47
median, **35**, 35–36
median smoothing, **88**
2-dimensional, 160
mesh, **171**
metadata, **175**
control, **175**
guide, **175**, 175–177
mistakes, making and admitting, 9
mode, **36**, 36–37
mother wavelet, **117**, 117–118
2-dimensional, 165
moving project
assumptions, 221
Gantt chart, 229, 230
goal, 220
network diagram, 227
task breakdown, 222, 224
task dependency, 226
work breakdown structure, 222, 223
mutually exclusive, **15**, 194
audience categories, 248
decision tree branches, 207
fundamental objectives, 194

negative frequency, **105**
negotiation, **212**, 212–215
BATNA, **213**, 214
individual value, claiming, **212**
issues, **213**, 214, 215

joint value, creating, **212**
reservation value, **213**, 214
ZOPA, **213**, 214
network diagram (in the project management context), **226**, 228
noise, *see* error, random
normalization, 109
Nyquist limit, **102**, 105
2-dimensional data sets and, 162
wavelets and, 120

objectives, 189, **192**
determining and structuring, 192–196
fundamental, **193**
negotiation issues, relation to, 213
new car example, 195–197
of analytic communication, 238
organization of, 194
swing weights and, 204
utility functions and, 201, 204
hierarchy, **194**
game theory and, 210
new car example, 195, 197
not proceeding beyond, 200
use in identifying alternatives, 198
means, **193**
new car example, 196
organization of, 194
negotiations and, 212, 213
overall fundamental, **194**
outcomes, **188**
attributes, relation to, 199
decision maker's appraisal using, 188
decision trees and, 207
negotiations and, 212
uncertainty in, 201
outliers, **30**

discarding, 54–55
elementary techniques and, 32
fitting and, 92
histograms and, 38
mean absolute deviation and, 46
mean, influence on, 34
median and, 36
running mean and, 86, 159
standard deviation, 46
variance and, 44

Parseval's theorem, **104**, 110
Pearson correlation coefficient, **134**, 168
period, **100**
phase, **100**
posterior probability, **180**
power spectrum, **104**, 110–111
precision, 30, **63**
preprocessing data, 175
prescriptive decision analysis, 21, 187–215
    origins of, 187
    see also decision, game theory, negotiation
previous work, reviewing, 6
prior probability, **179**, 182
prisoner's dilemma, 210–211
probability density function (pdf), **47**, 48
    repeated measurements and, 59
Program Evaluation and Review Technique (PERT), 223
project, **217**, 217–236
    assumptions, 221, 231
    charter, 220
    goal, **220**, 220–222
    hastening completion of, 231–232
    lead, **217**
        delegation and, 221
        metrics and, 221

speeding project completion and, 231
task uncertainty and, 225
typical management and, 218
management, 7, 191, **217**, 217–236
    alternative methodologies, 235
    as typically performed, 218
    methodology, 219–231
    of analytic projects, 232–235
milestones, **228**, 233
plan, **219**, 222, 234
roles, 221
scope, **220**, 222
see also critical path, Gantt chart, network diagram, tasks, and work breakdown structure
propagation of uncertainty, 65–68

quartiles, 36

range, **30**
ratios, **139**, 139–141
    2-dimensional data sets and, 169
    dynamic range and, 31
    interval scales and, 18
redundant coding, **248**, 252
regression, *see* fitting
related 1-dimensional data sets, 125–148
    correlation, 129–138
    correlation functions, 143–147
    examples, 127–129
    fitting, 142–143
    formalism, 125
    jointly measured, 126–127
        potential to introduce bias, 133
    ratios, 139–141
residuals, **51**, 98

resources, 191, **217**, 218, 228
restructuring data sets, 153–154
risk
    averse, **202**
    neutral, **202**, 208
    seeking, **202**
robust statistics, 32
running mean, **85**
    2-dimensional, 159

scales, **13**
    absolute, *see* scales, ratio
    categorical, **14**, 14–16, 20
        Bayesian analysis and, 180,
            181, 185
        communicating with color,
            253
        mode and, 37
        standard error of propor-
            tion, 70
    continuous, 20–21
        decision trees and, 207
        histogram and, 38
        mode and, 37
    discrete, 20–21
        decision trees and, 207
        histogram and, 38
    interval, **17**, 17–18
    nominal, *see* scales, categori-
        cal
    ordinal, **16**, 16–17, 221
        communicating with color,
            253
        confusion with ratio scales,
            19
        contrast with interval
            scales, 18
        median, use with, 36
        standard error of propor-
            tion, 70
    ratio, 17, **18**, 18–19
        age list and, 31
        decision trees and, 206
        height list and, 30
        negotiations and, 213

utility functions and, 201
scatter plots, **131**, 131–134, 166,
    169
sensitivity analysis, **205**
significance of results, 52–54
Simpson's rule, **50**
smoothing, **85**, 85–88
    2-dimensional, 159–160
    median, 88
    running mean, 85–87
sorting, 30
Spearman's ρ rank correlation co-
    efficient, 138
spectrum, *see* frequency analysis
stakeholders, **190**, 198
standard deviation, **45**, 45–46
    ordinal scales and, 17
standard error of proportion, **70**
stem and leaf diagrams, **41**, 41–42
swing weights, 205, **205**, 206, 213

tasks, **219**, 222–230
    dependency, **225**, 225–228,
        236
        finish to finish, **226**, 227,
            234
        finish to start, **225**, 227, 233
        start to finish, **226**
        start to start, **226**, 227
    effort-driven, **223**, 224, 230,
        232
    fixed-duration, **223**, 224, 230
    slack, **228**
    with uncertain durations,
        223–225, 230
temperature data set, **82**, 83
    annual pattern, 114
    fitting, 89
    increasing local contrast, 98
    Kendall τ correlation coeffi-
        cient, 138
    Lomb periodogram, 112, 113
    Nairobi, **128**, 130, 131
    Pearson correlation coeffi-
        cient, 136

ratio, 139, 141
  scatter plot, 132, 134
text mining, **176**, 176–178
  documents, **176**
  stemming, **177**
tradeoffs, 193
  and analytic communication, 238
  negotiations and, 212
  utility functions and, 201, 205
true positive, **179**, 180

uncertainty, 16, **58**, 57–73
  aleatory, **62**
  communicating, 199, 241
  decision outcomes and, 201
  epistemic, **62**
  histogram bins and, 38
  problems with usual methods, 71–73
  propagation of, 65–68
  task durations and, 223, 225, 230
  see also error
uncertainty in the mean, **68**, 68–71
unsharp masking, **99**
unstructured data, **174**, 173–186
  Bayesian techniques used with, 178–186
  extracting structure from, 174–176
  text as example of, 176–178
  see also metadata
utility function, **201**, 200–206
  decision trees and, 206, 207
  game theory and, 210

multi-attribute, 192, 197, 204, **204**, 205, 206
  negotiations and, 213, 214
  risk preferences and, 202, 208
  single-attribute, 204, **204**, 206

value-focused thinking, **192**
variance, **43**, 43–45
visualization, *see* data visualization
voltage data set, **79**
  discrete Fourier transform (DFT), 106, 107
  filtering, 115, 116
  fitting, 97
  power spectrum, 111
  running mean, 86, 87

wavelength, **100**
wavelet analysis, **115**, 115–121
  2-dimensional data sets and, 164–166
  cone of influence, **120**, 121
  mother wavelet, **117**, 117–118
    derivative of Gaussian (DOG), **117**, 119, 121
    Morlet, **117**, 119, 121
  power spectrum, **120**, 121
  significance and, 120
  wavelet transform, **119**, 119–121
weight list, *see* height list
window, **86**
  2-dimensional, 159
work breakdown structure, **222**, 223, 235

zone of influence, **86**, 88, 120

For Product Safety Concerns and Information please contact our EU
representative GPSR@taylorandfrancis.com
Taylor & Francis Verlag GmbH, Kaufingerstraße 24, 80331 München, Germany

www.ingramcontent.com/pod-product-compliance
Ingram Content Group UK Ltd.
Pitfield, Milton Keynes, MK11 3LW, UK
UKHW021619240425
457818UK00018B/642